Practice, Progress, and Proficiency in Sustainability (PPPS) Book Series

ISSN:2330-3271
EISSN:2330-328X

Editor-in-Chief: Ayman Batisha, International Sustainability Institute, Egypt

MISSION

In a world where traditional business practices are reconsidered and economic activity is performed in a global context, new areas of economic developments are recognized as the key enablers of wealth and income production. This knowledge of information technologies provides infrastructures, systems, and services towards sustainable development.

The **Practices, Progress, and Proficiency in Sustainability (PPPS) Book Series** focuses on the local and global challenges, business opportunities, and societal needs surrounding international collaboration and sustainable development of technology. This series brings together academics, researchers, entrepreneurs, policy makers and government officers aiming to contribute to the progress and proficiency in sustainability.

COVERAGE

- Socio-Economic
- Global Content and Knowledge Repositories
- ICT and knowledge for development
- Sustainable Development
- Strategic Management of IT
- Intellectual Capital
- Green Technology
- Environmental informatics
- Outsourcing
- Technological learning

IGI Global is currently accepting manuscripts for publication within this series. To submit a proposal for a volume in this series, please contact our Acquisition Editors at Acquisitions@igi-global.com or visit: http://www.igi-global.com/publish/.

The Practice, Progress, and Proficiency in Sustainability (PPPS) Book Series (ISSN 2330-3271) is published by IGI Global, 701 E. Chocolate Avenue, Hershey, PA 17033-1240, USA, www.igi-global.com. This series is composed of titles available for purchase individually; each title is edited to be contextually exclusive from any other title within the series. For pricing and ordering information please visit http://www.igi-global.com/book-series/practice-progress-proficiency-sustainability/73810. Postmaster: Send all address changes to above address.

Titles in this Series

For a list of additional titles in this series, please visit:
http://www.igi-global.com/book-series/practice-progress-proficiency-sustainability/73810

Handbook of Research on Bioeconomy and Economic Ecosystems
Ana Pego (Universidade Nova de Lisboa, Portugal)
Business Science Reference • © 2023 • 502pp • H/C (ISBN: 9781668488799) • US $295.00

Handbook of Research on Safe Disposal Methods of Municipal Solid Wastes for a Sustainable Environment
Prateek Srivastava (Chitkara University Institute of Engineering and Technology, Chitkara University, India) Durgaprasad Ramteke (Fibre and Particle Engineering, Faculty of Technology, University of Oulu, Finland) Ankush Kumar Bedyal (Department of Physics, School of Sciences, Cluster University of Jammu, India) Monika Gupta (Chitkara Business School, Chitkara University, India) and Jasminder Kaur Sandhu (Chitkara University Institute of Engineering and Technology, Chitkara University, India)
Engineering Science Reference • © 2023 • 414pp • H/C (ISBN: 9781668481172) • US $295.00

Hydrogen Fuel Cell Technology for Mobile Applications
Raluca Andreea Felseghi (Department of Mechanical Engineering, Faculty of Automotive, Mechatronics, and Mechanical Engineering, Technical University of Cluj-Napoca, Romania)
Engineering Science Reference • © 2023 • 313pp • H/C (ISBN: 9781668467213) • US $240.00

Energy Transition in the African Economy Post 2050
Olayinka Ohunakin (Covenant University, Nigeria)
Engineering Science Reference • © 2023 • 300pp • H/C (ISBN: 9781799886389) • US $215.00

Natural Resource Management Issues in Human-Influenced Landscapes
Dilini Hemachandra (Faculty of Agriculture, University of Peradeniya, Sri Lanka) and Jeevika Weerahewa (Faculty of Agriculture, University of Peradeniya, Sri Lanka)
Engineering Science Reference • © 2023 • 300pp • H/C (ISBN: 9781668470510) • US $225.00

Multidisciplinary Approaches to Sustainable Human Development
Vishal Jain (Sharda University, India) and Vinay Kukreja (Chitkara University, India)
Information Science Reference • © 2023 • 390pp • H/C (ISBN: 9781668482230) • US $225.00

For an entire list of titles in this series, please visit:
http://www.igi-global.com/book-series/practice-progress-proficiency-sustainability/73810

IGI Global
PUBLISHER of TIMELY KNOWLEDGE

701 East Chocolate Avenue, Hershey, PA 17033, USA
Tel: 717-533-8845 x100 • Fax: 717-533-8661
E-Mail: cust@igi-global.com • www.igi-global.com

List of Reviewers

Wolfgang Briglauer, *EcoAustria, Austria*
Aida M. Hanić, *Institute of Economic Sciences, Serbia*
Dler Salih Hasan, *Salahaddin University, Erbil, Iraq*
Maureen Hassall, *The University of Queensland, Australia*
Fredrick Ishengoma, *The University of Dodoma, Tanzania*
Ajeya Jha, Sikkim *Manipal Institute of Technology, India*
Bhim Jyoti, *College of Forestry, India*
Tajbia Karim, *Universiti Kebangsaan Malaysia, Malaysia*
Chen Jung Ku, *Universiti Malaysia Sabah, Malaysia*
Jitendra Kumar, *Sikkim Manipal Institute of Technology, India*
Yoke Lian Lau, *Universiti Malaysia Sabah, Malaysia*
Chek Kim Loi, *Universiti Malaysia Sabah, Malaysia*
Azilla Luthfiyyah, *Universitas Medan Area, Indonesia*
Zainal Rasyid Mahayuddin, *Universiti Kebangsaan, Malaysia*
Samrat Kumar Mukherjee, *Sikkim Manipal University, India*
Subhadeep Mukherjee, *Ramaiah University of Applied Sciences, India*
Ernahwatikah Nasir, *Universiti Malaysia Sabah, Malaysia*
Nicole Palan, *University of Graz, Austria*
Swarn Priya, *Symbiosis International University (Deemed), India*
Nikodem Rybak, *The University of Queensland, Australia*
Abhijit Sarkar, *SRM University, Sikkim, India*
Ahmad Budi Setiawan, *The National Research and Innovation Agency, Indonesia*
Ajay Kumar Singh, *DIT University, India*
Slavica M. Stevanović, *Institute of Economic Sciences, Serbia*

Wan Suryani, *Universitas Medan Area, Indonesia*
Pallavi Ghanshyala Vyas, *Jain University, India*
Tze Kie Yap, *Universiti Technology Mara Sarawak Malaysia*
Zi Hong Yong, *Universiti Malaysia Sabah, Malaysia*

Table of Contents

Detailed Table of Contents

Malaysia's digital gap index is a measure of disparities in internet access and utilization. The west Malaysian region has witnessed growth, but emotional intelligence and internet usage continue to present obstacles. To close the divide between those who have access to digital technologies and those who do not, the government must emphasize individual education. The implementation of 5G technology in Malaysia has the potential to reduce the digital divide and promote inclusiveness, thereby contributing to the growth of a sustainable and environmentally conscious economy. In east Malaysia, the implementation of sustainable development practices, community empowerment initiatives, and 5G infrastructure has the potential to reduce disparities and promote economic growth, ecological preservation, and social well-being. By providing equitable opportunities and promoting innovative economic and environmental progress, the government can effectively tap into the innovative potential and vitality of east Malaysians.

Public investment into the availability of ICT infrastructure has been studied for decades to assess its effect on economic development and productivity growth.

Whereas in the 1980s, this referred to the effects of telephones, since the 1990s it is the availability of internet connections, which is of great interest for politicians and policy makers alike. During the Covid-19 pandemic, social distancing has led to a transformation of work relations and the widespread introduction of work from home. Also, distance education as well as the increased importance of online streaming portals for leisure activities proved the necessity for fast broadband coverage and adoption. The authors investigate the degree of broadband coverage with a special focus on Austria and provide a survey on the economic benefits especially of high-speed internet connections for the economic development. The authors show to which extent Austria is using the full potential of fast and ultrafast internet connections and highlight the benefits for both firms and inhabitants in both rural and urban environments.

This study developed an information and communication technology evolution index (ICTEI) to explain the comparative performance of digitalization across 109 countries during 2010 – 2020. Composite Z-score and principal component analysis was used to create ICTEI. ICTEI was an integration of 16 different indicators associated with ICT access, ICT use, and ICT skills. Thereupon, ICTEI and per capita GDP were used as representative variables for digitalization and economic development, respectively to perceive a causal relationship among them. The empirical results revealed that economic development and digitalization have a positive and causal relationship. Trade openness, inflation, foreign direct investment, employment to population ratio, labour force participation rate, population growth, access to electricity, science and technological development, gross capital formation, and market exchange rate showed a diverse impact on economic development and digitalization in the different income groups countries.

In this chapter, the authors explore the theoretical and practical aspects of using text mining approaches supported by machine learning for the automatic interpretation of bulk literature on a contemporary issue—that of climate change risk analysis. The strengths, weaknesses, and opportunities associated with these approaches are investigated. Text mining provides a way to automate and enhance the analysis of text data. However, contrary to popular belief, text mining analysis is not a completely automated process. As with computer-assisted (or -aided) qualitative data analysis software (CAQDAS), it is an iterative method requiring input from a researcher with expert knowledge and a deliberate approach to the analysis. Given the heterogeneity that generally characterizes climate disclosures, the authors postulate that hybrid methodologies are ideal for analysing textual data related to climate change discourse. The authors also demonstrate that text mining is an open and evolving field, in the sense that it can be combined with other approaches to shed new light on the climate discourse.

Workplaces have been revolutionized by two factors: the rise of the Millennial age into the workforce and the emergence of online social networking tools as a means of connecting and collaborating with others. With the increasing penetration of technology in the form of artificial intelligence, IOT, 5G networks, and varied forms of social media, it is opportune for organizations to understand their impact on the current workforce. This research examines how Gen Y (millennials) use social media at work and their consequences on firms. Millennial employees seek organizational benefits such as increased knowledge sharing, work-related learning, employee engagement, and bridging social capital. The antecedents and consequences of social media use by Gen Y employees have been explained according to the uses and gratifications (U&G) theory.

Sign language is a medium of communication using gestures which is developed for people who suffer from auditory or verbal impairment. But sign language is usually not widely familiar to common people. For this reason, a sign language interpreter is essential for translating sign word representation to regular text or voice. Though there are some research works on American Sign Language interpretation, but it is inadequate in number for other languages. This chapter illustrates real time interpretation of medical emergency relevant Bangla sign words using machine learning. Considering medical emergency situations, a real time word level interpreter is designed, and the performance is evaluated. The designed deep neural network uses Mediapipe Holistic network for feature detection from human body in transitional movement and Long short term memory (LSTM) to predict the image sequences of each sign action. The designed system provides 99% model training accuracy and 92.22% test accuracy in real time detection.

Branded drug social media promotion (SMP) has increased dramatically in the recent era. The impact of this new phenomenon (although it is illegal in most countries) on patient behaviour has not been thoroughly studied. The purpose of this research is to learn about patients' perceptions of the impact of SMP on their behaviour. The sample size is 380 people. Pearson correlation and regression were used for analysis. The study clearly shows that SMP has an effect on patients' negative behaviour as well as the patient-physician relationship. The government and physicians should be conscious that proper guidelines are needed to guide patients. In addition, patients must understand that physicians and their expertise are critical to their treatment.

Technology is experiencing a revolution that is bringing about a new era successful in changing many tasks that have been manual processes into jobs with artificial intelligence (AI) that mimic human functions in solving cognitive problems and offer the same transformative potential to replace more potential tasks and activities. Human beings in various industrial, intellectual, and social applications. This era of AI technology has surprised us with new breakthroughs in learning and decision-making and raises new challenges and opportunities for sustainable innovation in green tourism marketing. The development of AI in Indonesia is more rapid as there are many internet users. Based on the results of a survey by the Internet Service Providers Association (APJII) in the 2022-2023 period, it was recorded that there were 215.63 million people. This number increased by 2.67% compared to the previous period in 2020-2021 of 210.03 million users.

This study focuses on investigating energy-aware resources and designing an energy-efficient cloud architecture. It aims to optimize the decentralization model and establish migration behavior to improve IT business operations using cloud computing. Several models were evaluated for performance metrics such as physical host use, power utilization, and service level agreement (SLA) misuse. These models include smart cloud optimization for resource configuration handling (SCORCH), V-MAN, decentralized energy-aware collaborative model (DEACM), efficient energy-aware resource management model (EEARMM), and energy and power-aware dynamic migration (EPADM). To handle resource configuration efficiently, the study employed V-MAN, DEACM, EEARMM, and EPADM within the framework of smart cloud optimization. Among these models, EPADM proved to be the most effective for resource utilization through dynamic VM migration. By leveraging EPADM, the cloud system can dynamically migrate virtual machines (VMs) to optimize resource allocation and energy consumption.

Preface

New advanced digital technologies and infrastructures will enable innovations and support the digital transition. Governments and companies must grasp this opportunity and take full advantage to create more inclusive and resilient economies and societies. Scientific collaboration networks and investment in technologies like AI, 5G/6G, blockchain, semiconductors and quantum will have a critical role (Almunawar et al., 2022; European Commission, 2023; Ordóñez et al., 2022 a,b).

The book offers innovative conceptual frameworks and theories, case studies and empirical studies to understand digital innovation and how transition in the digital age. Book chapters discuss how critical and enabling digital tools and solutions can boost the digital transformation of economies and societies and support the achievement of sustainable development goals (SDGs) by 2030. The outcomes of digital innovation must be in line with responsible research and innovation. The book will foster discussion on challenges, opportunities, and risks in the digital transformation among academics, researchers, industry players, policy makers and other relevant stakeholders. In conclusion the book will explore how these emerging topics can help to achieve and navigate the transition towards a circular, green and more sustainable economy.

OBJECTIVES OF THE BOOK

The book explores how advanced information technologies (artificial intelligence, machine learning, etc) can create digital tools and solutions to build more resilient, climate neutral and green economies and societies. The chapters of the book provide knowledge and insights on how digital tools and solutions and can impact the achievement of sustainable development goals (SDGs), with a special focus on the Asian region. In the case of the EU, European Comission's Horizon Europe Work Programmes promote 5G, AI, Internet of Things and digital transformation to boost the creation of jobs, build more resilient economies and societies and increase the inclusiveness of citizens.

Thanks to its innovative topics (5G, artificial intelligence, circular economy, green economy, climate neutral economy, inclusive growth) and approaches, geographical focus (international focus, with special attention to Asian region) and methodologies, the book will be a reference for students (undergraduate and postgraduate), academics, experts on IT, CEOs, policy makers and other relevant stakeholders around the world.

The book aims to be an international platform to foster dialogue among academics, researchers, decision makers, policy makers, and practitioners, share new theories and frameworks, research findings, and case studies, to enhance knowledge and scientific collaboration in digital innovation, business, green economy and inclusive growth. The book discusses how to boost circular, more sustainable and climate neutral economies and societies.

CONTENTS

The book offers a collection of 9 chapters.

Chapter 1, "A Catalyst for the 5G Era, a Green Economy, and Inclusive Growth: Closing Malaysia's Digital Divide," (by Yoke Lian Lau, Chek Kim Loi, Chen Jung Ku, Tze Kie Yap, Ernahwatikah Nasir and Zi Hong Yong) states that "Malaysia's Digital Gap Index is a measure of disparities in internet access and utilization. The west Malaysian region has witnessed growth, but emotional intelligence and internet usage continue to present obstacles. To close the divide between those who have access to digital technologies and those who do not, the government must emphasize individual education. The implementation of 5G technology in Malaysia has the potential to reduce the digital divide and promote inclusiveness, thereby contributing to the growth of a sustainable and environmentally conscious economy. In east Malaysia, the implementation of sustainable development practices, community empowerment initiatives, and 5G infrastructure has the potential to reduce disparities and promote economic growth, ecological preservation, and social well-being. By providing equitable opportunities and promoting innovative economic and environmental progress, the government can effectively tap into the innovative potential and vitality of east Malaysians".

Chapter 2, titled "Economic Benefits of High-Speed Broadband Coverage and Adoption" (by Wolfgang Briglauer and Nicole Palan) affirms that "public investment into the availability of ICT infrastructure has been studied for decades to assess its effect on economic development and productivity growth. Whereas in the 1980s, this referred to the effects of telephones, since the 1990s it is the availability of internet connections, which is of great interest for politicians and policy makers alike. During the Covid-19 pandemic, social distancing has led to a transformation of work relations and the widespread introduction of work from home. Also, distance education as well as the increased importance of online streaming portals for leisure activities proved the necessity for fast broadband coverage and adoption. The authors investigate the degree of broadband coverage with a special focus on Austria and provide a survey on the economic benefits especially of high-speed internet connections for the economic development. The authors show to which extent Austria is using the full potential of fast and ultrafast internet connections and highlight the benefits for both firms and inhabitants in both rural and urban environments".

Chapter 3, titled "Empirical Evidence from a Different Income Group Countries on the Interdependency between Digitalization and Economic Development" (by Ajay Kumar Singh and Bhim Jyoti) discusses a study that "developed an information and communication technology evolution index (ICTEI) to explain the comparative performance of digitalization across 109 countries during 2010 – 2020. Composite Z-score and principal component analysis was used to create ICTEI. ICTEI was an integration of 16 different indicators associated with ICT access, ICT use and ICT skills. Thereupon, ICTEI and per capita GDP were used as representative variables for digitalization and economic development, respectively to perceive a causal relationship among them. The empirical results exposed that economic development and digitalization have a positive and causal relationship. Trade openness, inflation, foreign direct investment, employment to population ratio, labour force participation rate, population growth, access to electricity, science & technological development, gross capital formation and market exchange rate showed a diverse impact on economic development and digitalization in the different income groups countries".

Chapter 4, titled "Machine Learning-Enhanced Text Mining as a Support Tool for Research on Climate Change: Theoretical and Technical Considerations" analyses "the theoretical and practical aspects of using text mining approaches supported by machine learning for the automatic interpretation of bulk literature on a contemporary issue—that of climate change risk analysis. The strengths, weaknesses and opportunities associated with these approaches are investigated. Text mining provides a way to automate and enhance the analysis of text data. However, contrary to popular belief, text mining analysis is not a completely automated process. As with computer-assisted (or -aided) qualitative data analysis software (CAQDAS), it is an iterative method requiring input from a researcher with expert knowledge and a deliberate approach to the analysis. Given the heterogeneity that generally characterizes climate disclosures, we postulate that hybrid methodologies are ideal for analysing textual data related to the climate change discourse. We also demonstrate that text mining is an open and evolving field, in the sense that it can be combined with other approaches to shed new light on the climate discourse".

Chapter 5, titled "Social Media and Gen Y at Work- The Uses and Gratifications of Technology" (by Pallavi Ghanshyala Vyas and Swarn Priya) discusses how "workplaces have been revolutionized by two factors: the rise of the Millennial age into the workforce and the emergence of online social networking tools as a means of connecting and collaborating with others. With the increasing penetration of technology in the form of Artificial Intelligence, IOT, 5G networks and varied forms of social media, it is opportune for organizations to understand their impact on the current workforce. This research examines how Gen Y (millennials) use social media at work and their consequences on firms. Millennial employees seek organizational benefits such as increased knowledge sharing, work-related learning, employee engagement, and bridging social capital. The antecedents and consequences of social media use by Gen Y employees have been explained according to the Uses and Gratifications (U&G) Theory".

Chapter 6, titled "Real-Time Bangla Sign Words Interpretation Uttered at Medical Emergencies" (by Tajbia Karim and Zainal Rasyid Mahayuddin) states that "sign language is a medium of communication using gestures which is developed for people who suffer from auditory or verbal impairment. But sign language is usually not widely familiar to common people. For this reason, a sign language interpreter is essential for translating sign word representation to regular text or voice. Though there are some research works on American Sign Language interpretation, but it is inadequate in number for other languages. This research paper illustrates real time interpretation of medical emergency relevant Bangla sign words using machine learning. Considering medical emergency situation, a real time word level interpreter is designed, and the performance is evaluated. The designed deep neural network uses Mediapipe Holistic network for feature detection from human body in transitional movement and long short term memory (LSTM) to predict the image sequences of each sign action. The designed system provides 99% model training accuracy and 92.22% test accuracy in real time detection".

Chapter 7, titled "An Empirical Study on the Impact of Social Media Promotion of Branded Drugs on the Health Behavior of Indian Patients" (by Samrat Kumar Mukherjee, Abhijit Sarkar, Subhadeep Mukherjee, Jitendra Kumar and Ajeya Jha) observes that "branded drug social media promotion (SMP) has increased dramatically in the recent era. The impact of this new phenomenon (although it is illegal in most countries) of promotion on patient behaviour has not been thoroughly studied. The purpose of this research is to learn about patients' perceptions of the impact of SMP on their behaviour. The sample size is 380 people. Pearson Correlation and regression were used for analysis. The study clearly shows that SMP has an effect on patients' negative behaviour as well as the patient-physician relationship. The government and physicians should be conscious that proper guidelines are needed to guide patients. In addition, patients must understand that physicians and their expertise are critical to their treatment".

Chapter 8, titled "Artificial Intelligence (AI) Opportunities and Challenges of Green Marketing Tourism" (by Wan Suryani and Azilla Luthfiyyah), notes that

"Technology is experiencing a revolution that is developing rapidly and bringing about a new era that has succeeded in changing many tasks that have been manual processes into jobs with artificial intelligence (AI) that mimic human functions in solving cognitive problems and offer the same transformative potential to replace potential tasks and activities. human beings in various industrial, intellectual and social applications. This era of AI technology has surprised us with new breakthroughs in learning and decision-making and raises new challenges and opportunities for sustainable innovation in green tourism marketing. The development of AI in Indonesia is more rapid as there are many internet users. Based on the results of a survey by the Internet Service Providers Association (APJII) in the 2022-2023 period, it was recorded that there were 215.63 million people. This number increased by 2.67% compared to the previous period in 2020-2021 of 210.03 million users".

Chapter 9, titled "A Study on Smart Energy Management Framework using Cloud Computing" (by Venkateswaran, Vidhya, Ayyannan, Manoj Chavan, Sekar and Boopathi) focuses on "energy-aware resources and designing an energy-efficient cloud architecture. It aims to optimize the decentralization model and establish migration behavior to improve IT business operations using cloud computing. Several models were evaluated for performance metrics such as Physical Host Use, Power Utilization, and Service Level Agreement (SLA) Misuse. These models include Smart Cloud Optimization for Resource Configuration Handling (SCORCH), V-MAN, Decentralized Energy-Aware Collaborative Model (DEACM), Efficient Energy-Aware Resource Management Model (EEARMM), and Energy and Power-Aware Dynamic Migration (EPADM). To handle resource configuration efficiently, the study employed V-MAN, DEACM, EEARMM, and EPADM within the framework of Smart Cloud Optimization. Among these models, EPADM proved to be the most effective for resource utilization through Dynamic VM Migration. By leveraging EPADM, the cloud system can dynamically migrate virtual machines (VMs) to optimize resource allocation and energy consumption".

Patricia Ordóñez de Pablos
University of Oviedo, Spain

REFERENCES

Almunawar, M. N., Islam, Z., & Ordóñez de Pablos, P. (2022). *Digital Transformation Management: Challenges and Futures in the Asian Digital Economy*. Routledge. doi:10.4324/9781003224532

European Commission. (2023). *Shaping Europe's digital future*. Europea. https://digital-strategy.ec.europa.eu/en

Ordóñez de Pablos, P., Zhang, X., Almunawar, M. N., & Labra Gayo, J. E. (2022). *Handbook of Research on Big Data, Green Growth, and Technology Disruption in Asian Companies and Societies*. IGI-Global. doi:10.4018/978-1-7998-8524-5

Ordóñez de Pablos, P., Zhang, X., & Almunawar, N. (2022b). *Green, Circular, and Digital Economies as Tools for Recovery and Sustainability*. IGI-Global.

Chapter 1
A Catalyst for the 5G Era, a Green Economy, and Inclusive Growth:
Closing Malaysia's Digital Divide

Yoke Lian Lau
https://orcid.org/0000-0002-7949-7683
Universiti Malaysia Sabah, Malaysia

Chek Kim Loi
Universiti Malaysia Sabah, Malaysia

Chen Jung Ku
Universiti Malaysia Sabah, Malaysia

Tze Kie Yap
Universiti Teknologi MARA, Malaysia

Ernahwatikah Nasir
Universiti Malaysia Sabah, Malaysia

Zi Hong Yong
Universiti Malaysia Sabah, Malaysia

ABSTRACT

Malaysia's digital gap index is a measure of disparities in internet access and utilization. The west Malaysian region has witnessed growth, but emotional intelligence and internet usage continue to present obstacles. To close the divide between those who have access to digital technologies and those who do not, the government must emphasize individual education. The implementation of 5G technology in Malaysia has the potential to reduce the digital divide and promote inclusiveness, thereby contributing to the growth of a sustainable and environmentally conscious economy. In east Malaysia, the implementation of sustainable development practices, community empowerment initiatives, and 5G infrastructure has the potential to reduce disparities and promote economic growth, ecological preservation, and social well-being. By providing equitable opportunities and promoting innovative economic and environmental progress, the government can effectively tap into the innovative potential and vitality of east Malaysians.

DOI: 10.4018/978-1-6684-8634-4.ch001

INTRODUCTION

The amalgamation of 5G, Green, and Sustainable Economy development is a subject matter that is gaining significance in both East (Sabah and Sarawak) and West Malaysia (Peninsular Malaysia). It is imperative to acknowledge the extant digital disparity and foster inclusiveness to guarantee that all localities and demographics can reap the advantages of 5G technology in propelling an environmentally conscious and sustainable economy. A digital divide exists between East and West Malaysia. In comparison to the more developed West Malaysia, East Malaysia exhibits a relative deficiency in internet connectivity, infrastructure development, and access to digital services. The existence of a digital divide results in an uneven distribution of opportunities for economic advancement and progress. The deployment of 5G technology offers a prospect to narrow the digital gap between East and West Malaysia. The extension of the 5G network infrastructure to remote and underserved regions in East Malaysia can lead to enhanced connectivity, access to digital services, and increased opportunities for economic participation, thereby benefiting the region. The promotion of green and sustainable practices has been more actively pursued in West Malaysia compared to East Malaysia. Malaysia has established formidable objectives to curtail carbon emissions and shift towards a low-carbon economic model. This encompasses the advocacy of sustainable practices such as the utilization of renewable energy sources, the implementation of energy-efficient measures, the adoption of sustainable transportation methods, and the support of ecologically conscious industries.

Throughout the Covid-19 outbreak, some Malaysians filmed their efforts to secure the best reliable internet connection. Others went to the tree, then to the village, and finally to the city (Muhammad, 2022). Veveonah, a student from a local university in Malaysia was chosen as the most renowned individual who climbed a tree to boost internet connectivity (Yap,2020). As a result of her video uploaded to her channel, the issue of East Malaysia's poor internet connectivity attracted international attention (Bernama, 2020b). Her account generated numerous news headlines (Jacobs & Subramaniam, 2020) and piqued the curiosity of Malaysian politicians and government officials (Doksil,2020). During the Malaysian education minister's visit to Sabah, she was very eager to meet Veveonah (Yahaya, 2020).

In an 11-minute and 20-second YouTube video released by Veveonah (Veveonah,2020), she did not criticize any Malaysian government official; rather, she described her difficulty locating an internet connection. A Malaysian government minister sarcastically referred her as "tree girl" and claimed she was lying because she did not have an exam at the time (Bunyan,2020). The Malaysian government issued an apology to Veveonah after her instructor revealed she wasn't lying (Yahaya, 2020).

The digital divide between East and West Malaysia has been a contentious issue for quite some time (Johari, 2020). Annually and at each general election, bridging the digital divide and shrinking the development gap between East and West Malaysia have been identified as top priority for addressing the issue (Editoron,2021). The digital divide between the East and West regions of Malaysia has not yet been overcome, despite the introduction of numerous project such as given incentives to promote personal computer use, acts, and regulations.

Tengku Datuk Seri Zafrul Tengku Abdul Aziz, the previous Malaysian Minister of Finance, announced on 7 October 2022, during the Malaysia Budget 2023 announcement that RM11.7 billion had been earmarked to East Malaysia's development goals (Minister of Finance, 2022). The was a yearly increase in funding given by the Malaysian government towards East Malaysia's development (Vinod,2021) to overcome the digital divide between East and West Malaysia. East Malaysia, comprises the states of Sabah and Sarawak, has not progressed considerably over time, despite its rich resources. Sabah and Sarawak remain the least developed and poorest states in Malaysia. This implies that numerous years of effort were squandered.

The digital divide is the difference between educational attainment and living standards (van de Werfhorst, Kessenich & Geven, 2020). During the Covid-19 pandemic, the digital divide between these two regions grew stark. Due to the widespread lack of WIFI access, most students in East Malaysia cannot access to online education. In East Malaysia, it is far more challenging for a student or employee who is assigned to work or study from home. This is because 2G, 3G, and 4G internet services are unavailable in the interiors of Sabah and Sarawak. During the Movement Control Order (MCO) in Malaysia, it is impossible to collect the data and information necessary to improve the living conditions of some Sabah Sarawak residents, particularly those residing in rural areas.

The Malaysia government is investigating the digital divide between East and West Malaysia because of the global growth of Covid-19. East Malaysia is sometimes portrayed as being in a state of decline, as being disadvantaged, and as having residents who continue to live in trees and commute by 'sampan'.

The ICT Roadmap and the Digital Malaysia project are among the strategies employed to ensure the achievement of the 10th Malaysia Plan. The 11th Malaysia Plan (11th MP) 2016–2020 aims to increase ICT utilization to enhance Malaysia's competitiveness and prosperity (Prime Minister Malaysia,2015).

Ghana (Baidoo-Anu, 2022), Pakistan (Jamil, 2021), India, China (Guo & Wan, 2022) but other countries also confront the same problem (Valk & Planojevic,2021; Kreitem, 2020; Kuráková, Vallušová & Marasová, 2021; Svongoro, 2022). Our analysis revealed that none of the respondents had negative comments regarding the government's efforts to close the digital gap. Respondents merely tell how they attempted to reduce the internet issues and do not indicate whose irresponsibility has led to the root issue.

From the onset of the Covid-19 pandemic to the present day, that is, three years have passed, and various experts held that the digital gap is a major issue that requires immediate government involvement. There have been numerous studies carried out relating to this aspect. This study attempts to address the dearth of studies evaluating the perspectives of Malaysian residents about the digital divide, particularly during the Covid-19 pandemic.

The implementation of 5G technology has the potential to expedite the advancement of intelligent urban centers in both East and West Malaysia. Smart cities utilize cutting-edge technologies, such as 5G, to optimize resource utilization, enhance public services, and establish ecologically sustainable urban ecosystems. Smart cities can leverage 5G technology to enable various applications such as intelligent transportation systems, efficient waste management, and smart energy management. Prioritizing the adoption of 5G technology in remote and underserved areas of East Malaysia is imperative to ensure inclusivity and bridge the digital divide. Strategic investments in infrastructure development, capacity building, and digital literacy programs are necessary to address this issue. Furthermore, it is imperative to establish a collaborative framework among governmental entities, private enterprises, and community organizations to foster an environment that facilitates the involvement of underrepresented populations and empowers them to leverage the advantages of 5G technology. To foster inclusiveness, it is recommended that initiatives aimed at community engagement and empowerment be implemented. The initiatives encompass the dissemination of knowledge regarding the advantages of 5G technology, implementation of educational and professional development schemes, and promotion of eco-friendly and sustainable business ventures at the community level. By engaging local communities in the decision-making processes and equipping them with the requisite knowledge and resources, the advantages of 5G technology can be extended to a broader demographic.

The research endeavors to address the subsequent research questions:

(i) To what extent does the digital divide index provide a precise reflection of the digital divide existing between the Eastern and Western regions of Malaysia?
(ii) What is the response of the residents in East Malaysia to the frequent internet disruptions?
(iii) What is the correlation between the digital divide and the four phases of the expansion of the digital realm?

The research objectives are as below:

(i) To calculate the digital divide index between East and West Malaysia.
(ii) To investigate the mindfulness attitude of respondents from the East Malaysia during internet interruptions as well as to determine the solution to closing the digital divide between East and West Malaysia.
(iii) To examine the potential economic benefits of 5G technology, environmentally friendly practices, and a sustainable economy, including job creation, novelty, and market expansion.

LITERATURE REVIEW

The Kementerian Komunikasi Dan Multimedia Malaysia (KKMM) assumes a crucial function in overseeing the advancement and execution of communication and multimedia policies, encompassing the realm of 5G technology (Kementerian Komunikasi Dan Multimedia Malaysia, 2018). The organisation engages in close collaboration with pertinent ministries and agencies to ensure the efficacious implementation of 5G infrastructure. The Malaysian Communications and Multimedia Commission (MCMC) functions as the governing body responsible for regulating the communications and multimedia industry within the nation of Malaysia. Its primary objective is to promote and facilitate the integration of 5G technology within the industry. Their responsibilities include spectrum planning, licensing, and monitoring the progress of 5G network deployment in the country (Malaysian Communications and Multimedia Commission, 2023).

The Ministry of Science, Technology, and Innovation (MOSTI) is tasked with spearheading the advancement and application of science, technology, and innovation in Malaysia. They engage in collaborative efforts with relevant stakeholders to advance sustainable and environmentally friendly technologies, including those related to 5G and sustainable economic practices (Ministry of Science, Technology, and Innovation, 2021). The Ministry of Energy and Natural Resources (KeTSA) prioritizes the advancement of sustainable energy development, energy efficiency, and conservation initiatives (Economic Planning Unit, 2022). They have a significant impact on promoting environmentally conscious and sustainable endeavors across diverse industries, including economics. Their efforts are directed towards mitigating carbon emissions and promoting the adoption of sustainable energy sources.

The Ministry of Environment and Water, commonly referred to as KASA, is responsible for safeguarding the environment and preserving its resources (Climate Change Adaptation Framework for Water Sectors, 2021). This includes promoting sustainable practices across various industries. They collaborate with relevant stakeholders to ensure the implementation of sustainable development initiatives. The Malaysian Green Technology and Climate Change Centre (MGTC) is a governmental entity operating under the purview of the Ministry of Science, Technology, and Innovation (MOSTI). Its primary objective is to encourage and streamline the advancement and integration of environmentally sustainable technologies. They focus on initiatives related to sustainable development, renewable energy, and environmental conservation (Shamsul Bahar, 2023).

The Malaysian Industry-Government Group for High Technology (MIGHT) is a governmental agency operating under the Ministry of Science, Technology, and Innovation (MOSTI). Its primary objective is to spearhead the advancement of high-technology and innovation in Malaysia. They engage in partnerships with various entities, including industry stakeholders, academic institutions, and governmental bodies, to foster sustainable economic expansion and technological progress (The Malaysian Industry-Government Group for High Technology (MIGHT), 2023).

Several 5G demonstration projects have been initiated in West Malaysia through collaboration between the Malaysian government and telecommunications companies. The objective of these projects is to demonstrate the potential utilization of 5G technology across diverse domains, including healthcare, manufacturing, agriculture, and smart cities. The Iskandar Malaysia development corridor in Johor is currently undergoing a transformation into a smart city, integrating 5G technology to enhance urban services, transport systems, and foster sustainable practices. The Smart Selangor Initiative, introduced by the Selangor state, aims to create intelligent urban areas within the state using advanced technologies such as 5G to augment public services, transportation, waste management, and energy efficiency. The Sabah Digital Economy Blueprint, initiated by the state government of Sabah, aims to transform Sabah into a hub for the digital economy by enhancing internet connectivity, fostering electronic commerce, and leveraging the potential of 5G technology for development initiatives. The Sarawak state government has implemented the Sarawak Digital Economy Strategy to propel economic expansion and promote digital transformation. The proposed approach encompasses various initiatives aimed at improving internet connectivity, setting up digital infrastructure, and leveraging cutting-edge technologies such as 5G to promote industrial expansion. The Rural Broadband Project has been initiated by the Malaysian government with the objective of providing internet connectivity to remote areas in Sabah and Sarawak. The project aims to address the issue of the digital divide and improve accessibility to digital services, including 5G technology. This information has been reported by TM One, Malay Mail, and Chung in 2023. Prior to embarking on the implementation of 5G technology and pursuing environmentally conscious and sustainable economic growth in Malaysia, it is imperative to first tackle the issue of the digital divide that exists in both the Eastern and Western regions of the country.

The term "digital divide" pertains to the disparity in internet accessibility, usage, and technological proficiency across diverse geographical areas and socioeconomic strata. This dichotomy has the potential to impede advancement and worsen pre-existing disparities. To promote equitable access to the advantages of 5G technology and sustainable economic progress, it is imperative to address the digital divide among the Malaysian population. The primary focus should be on ensuring equitable access to internet connectivity, enhancing digital literacy and competencies, and mitigating socioeconomic inequalities. The matter at hand pertains to tackling the obstacles encountered by rural regions, particularly those situated in the Eastern part of Malaysia, where access to internet connectivity is frequently restricted.

Diverse entities present distinct interpretations of the digital divide. The concept of "digital divide" has been defined by OECD (2001) and Dijk (2020) as the unequal distribution of Internet access and usage among individuals, households, businesses, and regions with differing socioeconomic backgrounds. The digital divide illustrates international and national differences. The so-called "digital divide" presents several difficulties.

According to Gonzales (2015) and Van Deursen & Van Dijk (2019) the initial phase of the digital divide consists of internet access and usage. Accessing technology is the first stage in the digital divide, which focuses on online access, is distinguished from the second level, which focuses on internet usage (Van Deursen & Van Dijk,2018,2019).

According to Scheerder et al. (2017), Van Deursen and Helsper (2015) as well as Ferreira et al. (2021), the third stage of the digital divide encompasses internet access, usage, and repercussions. For instance, Van Deursen and Helsper (2015) investigate economic, social, institutional, educational, and political outcomes. Scheerder et al. (2017) investigate all three factors (Internet skills, usage, and outcomes) in terms of socioeconomic, economic, social, cultural, personal, material, and motivational variables.

Different authors analyze the digital divide in distinctive ways, based on their distinct views. Chen (2017) examines digital inclusion policies, initiatives, and management strategies using examples of digital opportunity programs.

Dawood (2019) has undertaken research on the digital divide and rural poverty education in Malaysia's northern peninsular. Despite Malaysia's economic success in a variety of industries, poverty remains in several areas. Promoting ICT as the key to revitalizing and preserving regional communities is one strategy for reducing poverty. In Malaysia, however, there is a divide between urban and rural areas in terms of ICT availability. Even though numerous programs have been developed to bridge the digital gap, extra efforts are required, particularly in rural areas. The purpose of Dawood's study is to determine the extent to which ICT can support socioeconomic development in the rural regions of northern peninsular Malaysia.

The digital divide has long plagued Malaysians, particularly in East Malaysia. During the Covid-19 outbreak, when East Malaysians are unable to access the internet, they assume a vigilant disposition. This situation affects especially college students in East Malaysia.

By prioritizing initiatives to bridge the digital divide, Malaysia can lay a strong foundation for the successful implementation of 5G technology and sustainable economic practices. This approach will ensure that all Malaysians have equal opportunities to participate in the digital economy, access online services, and benefit from advancements in technology.

METHODS

The present research used Piaget's Cognitive Development Theory (Piaget,1964) to compare the actual human development to digital growth. Seeking, clicking, and downloading anything in the digital world constitutes exposure, like how a young child acquires sensorimotor abilities by searching, grabbing, and touching. The human digital world development stages comprise the following:

Stages 1: E-Readiness

E-readiness is not measured primarily by the number of computer servers, websites, and mobile phones rather, it promotes the utilization of digital technology. It is regarding a nation's technological readiness and capacity to engage in the global technology community. Typically, the level of maturity of a nation is determined by its ICT infrastructure and its population's ability to utilize ICT for sustainable development (Hanafizadeh, Hanafizadeh & Bohlin, 2013).

From our perspective, e-readiness works as a barometer in the digital world. E-readiness indicates how far a person can advance in the digital world (Hanafizadeh, Hanafizadeh & Saghaei, 2013). The sensory-motor development of a person is linked to their ability to travel. Connecting to the internet is the basic step required to enter the world of the internet. In addition to speed and coverage, internet data determines the signal's range. Our team illustrates the concept by comparing physical and digital letter delivery.

Sensory-motor connection enables a person to feel (Schenkman,2021) a letter and open an email address (e readiness is required to access an email address). In the real world, a person touches an actual letter and then writes on its blank paper. In the digital world, however, an email may be typed on a blank page within an email.

Stage 2: E-Concept Preoperational

A person must learn how to write a letter, how to seal it, and where to mail it before sending one. One must learn where to click "new" to generate a new email, where to enter the email address, and where the send button is located.

Stage 3: E-Knowledge Concrete Operational

It is to simplify the procedure, one must regularly practice sending a letter to avoid missing any steps, which could result in a failed postal transmission. In the digital age, email writing is an art that must be perfected via ongoing practice.

A person can speak, walk, move, converse, etc., after obtaining other real-world abilities. Moreover, if a person is born with sensory motor deficits, he or she can nonetheless develop in the actual world, albeit with limited capabilities. Waiting for the email page to load without Internet access will result in nothing being displayed. The sensory-motor impairment will enhance a person's problem-solving skills in the real world, whereas the e-readiness issue will bring difficulties in virtual world development.

If a person has not gone through the complete development process in the digital world, he may be at a disadvantage in the digital world (Heimann, Bus, Barr, 2021; Klaus,2017), even if he is incredibly successful in the physical world. A person may lack the ability to compose an email.

Stage 4: E-World Formal Operational

If a person is accustomed to the digital world, for example, he will apply the concept that to become a member of a mailbox, he or she must sign up as a member. If a person is searching on another webpage and wants to become a member online, he or she will look for the word "sign up" on the relevant pages. For a person to communicate or perform his or her job in the digital world, he or she must complete all this training and achieve these outcomes.

The stages of human evolution in the physical and digital worlds are presented in *Table 1*.

Only after experiencing all four stages of growth in real life could someone have a similar perspective on development. If a person does not complete all four phases, communication difficulties may result, as is the situation with some autistic individuals, who typically face problems in sensory-motor problems.

In the digital age, a person lacking essential computer abilities may be termed incompetent or incapacitated. In the digital environment, a person's development could be terminated at any time. Due to a lack of internet connectivity, the application

Table 1. The Piaget's cognitive development theory (1964) in real world and the digital world development theory

Real World	Stages	Digital World
Sensory Motor	1	E-readiness[1]
Preoperational	2	E-concept preoperational
Concrete operational	3	E-knowledge concrete operational
Formal operational	4	E-world formal operational

[1]In the digital world, E-readiness serves as a human sensorimotor system.

is in-accessible in some places with limited Internet access. Some individuals may abandon the topic at the second stage if they are unable to learn further. Regional disparities in the digital divide may be attributable to varying rates of digital growth.

The first, second, and third levels of development are formal operational preparation levels that enable individuals to conduct official operations in a physical or virtual environment. The digital divide currently consists of three main layers. Research on e-readiness focuses mostly on the digital divide at level 1, which consists of various components. Internet quality research is the second level of the digital divide, which refers to disparities in how individuals utilize digital resources and is the outcome of the need to expand the idea of the digital divide. In the digital age, some individuals have already attained two or three stages of human growth. The third phase of the digital divide focuses on the real-world consequences of disparities in access to and usage of digital resources, some researchers may see the characteristics of the digital divide levels differently, such as Danial, who classifies digital emotion as the second digital divide level. EQ and emotions when engaging in formal academic pursuits online. The third phase of the digital divide study may include studies on robot development, cyberbullying, smart us-age, and online proficiency.

The research focuses on sociodemographic and the normal association between internet access and the earliest stage of the digital divide. Most places with a significant digital divide are in stage two of development, or even worse, are uninformed of the existence of internet-connected gadgets. Most of the research on the first and second digital divides focuses on regional disparities in internet access and ICT infrastructure. In industrialized nations with advanced ICT and internet infrastructure, the third level of the digital divide is routinely investigated to get a deeper comprehension of intelligent concept creation, robotics, internet-related emotions, etc.

The *Table 2* shows the Human development in the digital world and its relations to digital divide levels.

About the research methods, the present study examined four dimensions using (i) questionnaire and (ii) WhatsApp regular internet connection report from Universiti

Table 2. Human development in the digital world and its relation to digital divide levels

Stages	Digital Divide Levels	Issues
E-readiness	Accessibility to internet and various ICT	WIFI Coverage
E-concept Preoperational	Differences in digital usage	Outdated digital usage.
E-Knowledge Operational	Differences in digital usage	Cyberbully, skill training
E-World Formal Operational	Digital Emotional	Digital EQ

Malaysia Sabah (UMS) students. The questionnaire has three sections: personal information, the 13-questions about Covid-19, and a 10-questions internet survey. Only three dimensions were considered in this investigation. They are background checks, Internet familiarity, and current affairs discussions on the COVID-19 outbreak.

The first section of the questionnaire seeks to evaluate the initial stage of the digital divide, which is always associated with the socioeconomic profile of a community. For the second component of the questionnaire, Covid-19 information must be researched. During MCO, most of the Covid-19-related information was disseminated through the internet and television. The higher a respondent's score on this component, the more efficiently they can retrieve Internet-based information. This pertains to the second phase of the digital divide, which emphasizes various digital usage styles. They cannot access news online without a mobile device. The final component of the questionnaire asks respondents to estimate their online time and Internet usage. This question is part of the third level of the digital divide and asks respondents to evaluate their usage's mood and utility.

Our team analyses the digital divide between East and West Malaysia using the Gini coefficient. The primary advantage of the Gini coefficient is that it is a ratio-based measure of inequality, as opposed to a variable unrepresentative of most of the population, such as per capita income or gross domestic product. It can be used to compare income distributions across different population sectors and countries, such as how the Gini coefficient for urban areas differs from that of rural areas in numerous nations (Fujita, 2021; Mukhopadhyay & Sengupta, 2021).). Our team applies the Gini coefficient to our analysis and calculates the digital divide index between East and West Malaysia using this formula.

The calculation formula is as follows:

Digital Divide Index (DDI)=

A= Total Indicator East Malaysia (Total Indicator Sabah+ Total Indicator Sarawak)

B=Total Indicator West Malaysia

Total Indicator Sabah plus Total Indicator Sarawak is the formula for Total Indicator East Malaysia. The total indicator for Sabah will be derived from Sabahans' responses to questionnaire (parts B and C), whereas the total indicator for Sarawak, will be derived from Sarawakians' responses to questionnaire (parts B and C).

The total indicator for Sabah the formula is as below:

Total indicator Sabah= [(Indicator Sabah part Bx0.5)
+(Indicator Sabah part Cx0.5)] x0.5

And total indicator for Sarawak the formula is as below:

Total indicator Sarawak= [(Indicator Sarawak part Bx0.5)
+(Indicator Sarawak part Cx0.5)] x0.5

Total indicator West Malaysia= (Total indicator West Malaysia Part Bx0.5)
+ Total indicator West part Cx0.5)

The hypothesis of the Gini index:

a. <0.2 means the gap between the two areas is low.
b. 0.2-0.29 means the gap between the two areas is medium.
c. 0.3-0.39 means the gap between the two areas is medium up.
d. 0.4-0.59 means the gap between the two areas is high.
e. 0.6 and above means the gap between the two areas is very high

During MCO, our research team collected approximately 436 questionnaires using Google Forms. Our target respondents can be divided into three categories. Sabahans who have resided in Sabah for at least fifteen years. Second, Sarawakians must have spent at least 15 years in Sarawak. West Malaysia citizens must have stayed in West Malaysia for at least 15 years and were born in West Malaysia. This is to ensure that the respondents are familiar with their state and retain state-specific behaviors. Among the 436 questions, we focused solely on the 20-25-year-old age group. In addition, after screening, they qualify for Sabahan (46), Sarawakian (51), and West Malaysian (58).

RESULTS AND DISCUSSION

Table 3 displays indicators of each question and its weightage for questionnaire for part B.

Table 3. Indicators of each question and its weightage for questionnaire (B)

Q*	Respondents Sabahan 46 people				Respondents Sarawakian 51 people				Respondents from West Malaysia 58 people			
	Score	Total Mark	Weightage 1/13	Indicator Weightage X0.5	Score	Total Mark	Weightage 1/13	Indicator Weightage X0.5	Score	Total Mark	Weightage 1/13	Indicator Weightage X1
1	14	46	0.023	0.007	20	51	0.030	0.015	53	58	0.070	0.070
2	40	46	0.066	0.022	50	51	0.075	0.037	56	58	0.074	0.074
3	15	46	0.025	0.012	15	51	0.022	0.011	55	58	0.072	0.072
4	46	46	0.076	0.038	51	51	0.076	0.038	58	58	0.076	0.076
5	39	46	0.065	0.032	45	51	0.067	0.033	53	58	0.070	0.070
6	40	46	0.066	0.033	49	51	0.073	0.036	56	58	0.074	0.074
7	16	46	0.026	0.013	29	51	0.043	0.021	57	58	0.075	0.075
8	18	46	0.030	0.015	28	51	0.042	0.021	58	58	0.076	0.076
9	18	46	0.030	0.015	22	51	0.033	0.016	55	58	0.072	0.072
10	6	46	0.010	0.005	14	51	0.021	0.010	20	58	0.026	0.026
11	20	46	0.033	0.016	17	51	0.025	0.012	28	58	0.037	0.037
12	40	46	0.066	0.033	49	51	0.073	0.036	58	58	0.076	0.076
13	43	46	0.071	0.035	50	51	0.075	0.037	58	58	0.076	0.076
Total Indicator				0.276				0.323				0.874
0.276x0.5=				0.013	0.323x0.5=			0.161	0.874x0.5=			0.437

Table 4 displays indicators of each question and its weightage for questionnaire for part C.

Based on Table 3 and Table 4 given indicators, the Total indicator East Malaysia is (0.013+0.059) + (0.161+0.142) =0.375, and the Total indicator West Malaysia is 0.437+0.349=0.786.

Digital Divide Index (DDI)= 0.375/ 0.375+0.786=0.322

The hypothesis of the Gini index:

a. <0.2 means the gap between the two areas is low.
b. 0.2-0.29 means the gap between the two areas is medium.
c. 0.3-0.39 means the gap between the two areas is medium-upper.
d. 0.4-0.59 means the gap between the two areas is high.
e. 0.6 and above means the gap between the two areas is very high

Table 4. Indicators of each question and its weightage for questionnaire (C)

Q*	Respondents Sabahan 46 people				Respondents Sarawakian 51 people				Respondents from West Malaysia 58 people			
	Score	Total Mark	Weightage 1/10	Indicator Weightage X0.5	Score	Total Mark	Weightage 1/10	Indicator Weightage X0.5	Score	Total Mark	Weightage 1/10	Indicator Weightage X 1
1	61	184	0.033	0.016	145	204	0.071	0.035	183	232	0.078	0.078
2	110	368	0.029	0.014	307	408	0.075	0.037	360	464	0.077	0.077
3	42	92	0.045	0.022	98	102	0.096	0.048	106	116	0.091	0.091
4	10	92	0.010	0.005	20	102	0.019	0.009	39	116	0.033	0.033
5	13	92	0.014	0.007	39	102	0.038	0.019	42	116	0.036	0.036
6	15	46	0.032	0.016	40	51	0.078	0.039	55	58	0.094	0.094
7	17	46	0.036	0.018	32	51	0.062	0.031	50	58	0.086	0.086
8	8	46	0.017	0.008	32	51	0.062	0.031	37	58	0.063	0.063
9	3	46	0.006	0.003	13	51	0.025	0.012	39	58	0.067	0.067
10	9	46	0.019	0.009	25	51	0.049	0.024	43	58	0.074	0.074
Total Indicator				0.118				0.285				0.699
	0.118x0.5=			0.059	0.285x0.5=			0.142	0.699x0.5			0.349

Q* refers to questions.

According to the digital divide index, East and West Malaysia are in the middle-to-upper group. The digital gap index reveals that East and West Malaysia continue to have discrepancies in internet accessibility and usage rates. Adoption of the e-concept and e-knowledge in East Malaysia is insufficient, as is its accessibility. West Malaysia is well developed till the issue of EQ and emotions when using the internet is broadened, whereas in East Malaysia there are still areas without WIFI access.

As for the 'WhatsApp regular internet connection reports ', this regular internet connection reports include 93 Universiti Malaysia Sabah students. Prior to enrolling in an online Mandarin lesson, they must submit their current internet connection status and report it through WhatsApp.

Among the obstacles faced by the students were flooding, blackouts, and insufficient internet connectivity. The students' proposed alternatives include going to the city to find an Internet connection and heading to their part-time jobs to connect to the Internet. Approximately 70% of students had experienced at least 10 times or more during a year (second and third level) of online mandarin instruction. Approximately 90% of users experience at least ten instances of internet loss. According to the report, all of them have shown the thoughtful attitude of accepting the facts and working to resolve the issue; none of them condemned the government for not developing internet infrastructure. Some of the student responses are listed below:

"I'm sorry, I can't connect to the internet, I can't attend the lesson today, and the battery on my laptop is going to die..." (Respondent 0584 at 19.14pm on 9 May 2022)

"Miss, I apologies, but I am unable to attend class today due to a blackout in my home from day to night and a weak internet connection..."(Respondent 4694 27 December 2021 11.53am)

"Miss, I'm sorry, but I'm unable to attend the online class today due to a power outage at home and a poor internet connection. If I have a phone line, it's just E." (Respondent 0954 06 December 2021 11.40am)

Most respondents regret that the loss of internet caused them to be absent or unable to complete the online exam paper on time. In general, their comments demonstrate their eagerness to enroll in the online course, despite the numerous obstacles they must overcome.

Based on the above findings, the present study answers the research questions formulated in this study as follows:

(i) How accurately does the digital divide index reflect the digital divide between East and West Malaysia?

Now is a suitable time for digital reasoning. We concurred that immediate action is essential to compel the Malaysian government to examine strategies to reduce the digital divide index between the country's two regions. The Malaysian government should maybe adopt the trend of digitally boosting the value of systems.

(ii) How have residents in East Malaysia responded to the regular internet outages?

According to the UMS students' WhatsApp regular internet connection reports, the majority of them were experiencing flooding, blackouts, and poor internet service. Water shortages, etc., they still maintained a mindful attitude and didn't condemn the government for not doing their task well. Therefore, their perspective on the digital divide gap was non-judgmental, patient, and accepting of the reality. During the Covid-19 pandemic, they did not assign a negative label to the authorities who did not perform their duties flawlessly and caused them to fall victim (Desormeau, Walsh & Segal, 2018; Reibel & McCown,2019; Lahtinen & Salmivalli, 2020) to the digital gap. Regarding internet connectivity concerns, they were very patient and diligent in their pursuit of a superior solution. They also understood and treated the reality that there was a gap between East and West Malaysia with calmness.

(iii) What is the relationship between the digital divide and the four stages of the growth of the digital world?

According to Surianshah's (2021) study, a student's low online learning attendance has a positive and statistically significant effect on their impression of the effectiveness of online learning. The study employed primary data collected from secondary school students in rural Sabah, Malaysia. Notably, the significant influence on student perspective persisted after the reverse causality effect between online learning attendance and online learning had been addressed. However, low online attendance has little effect on students' opinions. This study also revealed that the lack of digital devices at home has a significant impact on a student's decision to work part-time in order to enroll in online courses. These statistics indicate that rural students continue to face considerable digital access barriers. The research conducted by Surianshah(2021) demonstrates that Sabah remains at digital divide level 1. The internet use requirements are still insufficient. Human growth in the digital environment was halted at the e-readiness stage.

According to Ayob, Aziz & Ayob (2022), from the perspective of continued development, however, the advent of digital services such as the Internet may increase the area's productivity. Stable and rapid Internet access is now recognised as a need. If enough internet access is made available in rural areas, tourism, food, and traditional goods sectors, among others, may be able to flourish there. Foreign guests require internet access for a variety of reasons, including study, e-hailing service booking, and remote business. The provision of Internet connection in rural areas is not a priority from the perspective of population density. However, when viewed through the lens of long-term investment, these industries have the potential to increase economic production. This has the potential to eliminate the digital divide between rural and urban areas.

Based on our finding, stages 1 of e-readiness involve the preparation of facilities and internet infrastructure. This relates to the government's capacity to prepare citizens for internet-based work. The better the facilities, the higher the phases of digital human development that citizens can attain. We must adequately prepare pupils until at lEast the stages 3 of human development in digital world, after which they will be able to freely choose and value the most effective learning platform. All students who can identify their preferred platform have reached the final stage of human growth in the digital world. Online education may be the future instructional approach. According to Amir and Borhan (2022), students appreciated synchronous learning sessions where WhatsApp, Google Meet, and Google Classroom were used to facilitate contact between the professor and each student during critique sessions or tutorials.

According to the research of Bahar, Wahab, & Ahmad (2020), the obstacles of online teaching and learning include inadequate real-time communications, technological incompetence, a low rate of engagement and interaction, and a lack of awareness of the characteristics and attitudes of learners. In addition, their research offers prospective strategies for overcoming hurdles and improving online training and delivery.

Our research finding shows that if a student had not reached phases 2 (E-concept preoperational) and 3 (E-knowledge concrete operational), he or she could become confused when confronted with the issue during an online session, according to our assessment. Those who possess the necessary competence and knowledge of the digital world may not experience worry when the problems in the digital world develop. Furthermore, it is common knowledge that such situations are commonplace in the digital age. This highlights the importance of human development phases inside the digital environment.

Bahruddin and Ramadhanti Febriani (2020) note that the online Arabic learning process is influenced by several factors, including student internal motivation, lecturer supervision, appropriate facilities, and access to materials. Therefore, the shortcomings of the techniques can be utilised to improve online Arabic education. Students remain motivated to learn, particularly during COVID-19, by mixing all learning components in a holistic and diverse manner. According to the findings of this study, the importance of teacher competency in the pedagogical domain is to provide students with understanding, organise and implement learning, and develop students' different potentials. Consequently, the pedagogical skill of a professional lecture can encourage a dialogical and participatory learning environment. Additionally, it is a complex learning concern. It also affects the success rate of the process and student learning outcomes. This study examines the formal operation of the E-world by pupils. All the traits highlighted by Bahruddin and Ramadhanti Febriani (2020) are those exhibited by the user in the digital realm.

Malaysia has initiated multiple 5G demonstration projects to exhibit the prospective applications of the technology. The Langkawi 5G initiative incorporated intelligent tourism solutions, including virtual reality (VR) tourism experiences and intelligent parking systems, (Bernama, 2020a) to augment the tourist experience while advocating for sustainable tourism practices. As part of its digital transformation agenda, Malaysia has set its sights on the development of multiple digital cities powered by 5G technology. Metropolitan areas such as Kuala Lumpur and Cyberjaya are poised to utilize 5G technology to establish sustainable and intelligent urban ecosystems. This will involve the incorporation of advanced features such as smart transportation, energy management, and efficient public services.

The Malaysian government has implemented sustainable energy development initiatives, including the Large-Scale Solar (LSS) projects (SEDA, 2023a) and the Net Energy Metering (NEM) program (SEDA, 2023b). These initiatives facilitate the production of sustainable energy and empower both individuals and enterprises to engage in prosumers by generating their own environmentally friendly energy. The government of Malaysia has been proactively engaged in the deployment of smart grid technologies with the aim of enhancing the effectiveness and dependability of the electricity grid. This endeavour entails the incorporation of cutting-edge technologies, such as 5G, to facilitate instantaneous monitoring, energy optimization, and demand-response mechanisms.

Malaysia boasts a diverse range of green building projects and initiatives that are geared towards mitigating the environmental footprint of buildings. The Malaysian government is presently involved in advocating for sustainable transportation options, including electric vehicles (EVs) and public transportation systems (Afiq, 2023). GTRADE is an initiative that places emphasis on the electrification of rail systems, with the objective of reducing emissions and improving transport efficiency (MIGHT, 2014).

Malaysia is presently conducting an inquiry into the adoption of agroecology and sustainable agriculture approaches to enhance food security while concurrently minimising any adverse environmental impacts. The initiatives involve the advancement of organic farming, agroforestry, and precision agriculture technologies that employ data analytics and IoT to enhance resource management. The endeavours highlight Malaysia's commitment to leveraging 5G technology and sustainable methodologies to drive economic progress, ecological conservation, and societal well-being.

It is imperative to ensure that the advantages of said policies and projects are extended to all areas of Malaysia, encompassing the Eastern region. The assessment of the extent to which policies and projects are benefiting companies in these fields, as well as the evaluation of whether companies in East Malaysia equitable opportunities must participate, is significant.

To attain the intended results, it is imperative to oversee and ascertain the proper and efficient execution of all corporate projects. This entails not only the dissemination of positive figures and results, but also the proactive involvement with the corporations in question. The provision of essential support, guidance, and resources is crucial in facilitating the awareness and complete involvement of enterprises in East Malaysia in said initiatives. Through this approach, Malaysia has the potential to cultivate an environment of inclusivity, promote equitable economic growth, and establish a more sustainable future for all regions.

CONCLUSION

The present study has the following implications:

(i) Calculate the digital divide index between East and West Malaysia

When the digital divide index decreases, the digital divide narrows. Various scientists have developed the concept of the digital divide index to quantify the digital gap. Various academics measure the digital divide index utilizing a variety of formulas and concepts. By developing a future formula, the government may need to take additional measures to investigate the digital divide between East and West Malaysia. The research serves as a rapid instrument for collecting and analyzing essential data to provide the government with a comprehensible index and to take further action. If another emergency like Covid-19 were to occur, the dependent variable could be modified to reflect the current state. Even if individuals in the target research locations are unable to respond to the survey, we can target individuals with comparable digital behaviors who reside in internet-accessible areas and request that they submit study-related data.

(ii) To investigate the mindfulness attitude of respondents from East Malaysia during internet interruptions

East Malaysians always stay in mindfulness. Before the nation can achieve a higher level of digital world development, the government must improve the digital skills of its population to bridge the digital divide. According to Ambali and Bakar (2014), information and communication technologies have the potential to dramatically enhance the delivery of public services; nevertheless, user behaviour is one of the greatest obstacles to the successful adoption of ICT systems. Before fully integrating ICT into the public domain, it is necessary to foster user approval. ICT acceptance and application in the Malaysian Public Sector investigates in depth the disparities in ICT theory, user attitudes, and the underlying reasons that limit the full adoption of ICT systems to address the obstacles associated with the implementation of practical ICT systems in the public sector.

The majority of those affected by the digital divide are aware of the issue, which enables the government to participate in a long-term programme without feeling compelled to achieve its goal. In times of relative calm, the Malaysian government should work fast to equalise East and West Malaysia in recognition of Malaysians' tolerant and accommodating temperament. Concerns about the digital gap have not been resolved in recent years in part because East Malaysians are so forgiving, less violent, constantly minimalistic, nonaggressive, and accepting of injustice. As a result, many efforts designed to bridge the digital divide in East Malaysia are insufficient or useless since nobody will complain because East Malaysians do not have a culture of complaining.

Our houses or rooms have been placed under lockdown. If a similar or same occurrence occurs in the future, there will be further opportunities to get similar experiences. Existing lockdown experiences may serve as a template for future lockdowns or the enforcement of lockdowns to safeguard human life and routine. If we ever lost control or authority in the physical world, we could only rely on the digital world, whose functional ecosystem would enable us to save the physical world. In the digital era, ensuring that a person recovers and completes the development process may be the most important duty for all global governments. This is a crucial preparation for their citizens to face the unknown challenge one day when we have all lost control of our rights (including the ability to travel anywhere) in the physical world, so that we can at least continue to communicate in the digital world and cooperate with the authority when necessary to save back our physical world.

After Covid-19, we realized that the digital divide is not simply a matter of inequity, but also of our ability to survive in the real world. How do rural residents who lack access to reliable WIFI receive the information necessary to learn how to deal with Covid-19? Ensure that the evolution of our physical and digital worlds is parallel. Everyone should have the right to a comprehensive education in the digital world and to experience the complete development process in both the physical and digital worlds. Make the right to participate in the digital world's development process a new human right and ensure that no one is left behind. The physical and digital worlds are now intertwined, and an individual may be disadvantaged without either.

In both the physical and digital realms, young Malaysians serve the new ideology. Those who originated from less developed phases are mindful, patient, realistic, and eager to advance. This allows the government considerable time to build the necessary digital infrastructure to support a big user base.

(iii) To examine the potential economic benefits of 5G technology, environmentally friendly practices, and a sustainable economy, including job creation, novelty, and market expansion.

By addressing the digital divide and promoting inclusivity, Malaysia can harness the potential of 5G technology to drive a green and sustainable economy. The deployment of 5G infrastructure in East Malaysia, along with sustainable development initiatives and community empowerment, will help bridge the gap and create opportunities for economic growth, environmental conservation, and social well-being. It is important to recognize the positive thinking and potential of the people in East Malaysia. For the efforts to move towards a 5G, green, and sustainable economy to succeed, the government must ensure that equal resources and opportunities are given to East Malaysia. By providing equal support and opportunities, the government can tap into the talent and enthusiasm of the people in East Malaysia, fostering innovation and driving economic and environmental progress in the region.

REFERENCES

Afiq, H. (2023, April 13). Is Malaysia ready for commercial EVs? *The Malaysian Reserve*. https://themalaysianreserve.com/2023/04/13/is-malaysia-ready-for-commercial-evs/. https://themalaysianreserve.com/2023/04/13/is-malaysia-ready-for-commercial-evs/

Ambali, A. R., & Bakar, A. N. (2014). *Ict adoption and application in the Malaysian public sector*. Information Science Reference.

Amir, A. F., & Borhan, R. (2022). Students' perception and preference towards open and distance learning (Odl) during COVID-19 pandemic: A case study of studio-based learning. *Malaysian Journal of Sustainable Environment*, *9*(1), 285. doi:10.24191/myse.v9i1.17304

Ayob, N. H., Aziz, M. A., & Ayob, N. A. (2022). Bridging the digital divide: Innovation policy and implementation in Malaysia. *International Journal of Academic Research in Business & Social Sciences*, *12*(8). doi:10.6007/IJARBSS/v12-i8/14554

Bahar, N., Wahab, S. N., & Ahmad, N. D. (2020). Understanding challenges faced in online teaching and learning among Malaysian universities' instructors during COVID-19 pandemic. *2020 Sixth International Conference on e-Learning (econf)*. IEEE. 10.1109/econf51404.2020.9385474

Bahruddin, U., & Ramadhanti F. S. (2020). Student's perceptions of Arabic online learning during COVID-19 emergency. *Journal for the Education of Gifted Young Scientists*, 1483-1492. doi:10.17478/jegys.763705

Baidoo-Anu, D. (2022). Digital divide in education: Exploring undergraduate Students' Online learning experiences in Ghana during the COVID-19 pandemic. *AERA, 2022*. Advance online publication. doi:10.3102/IP.22.1885643

Bernama. (2020a). *Bernama: 12 Jan 2020: TM propelling smart tourism through 5G in Langkawi*. Official Portal of Ministry Of Communications And Digital. https://www.kkd.gov.my/en/public/news/16284-bernama-12-jan-2020-tm-propelling-smart-tourism-through-5g-in-langkawi

Bernama. (2020b). UMS student's girl in taking exam on a tree won viewers' hearts. *Astroawani.* https://www.astroawani.com/berita-malaysia/ums-students-grit-taking-exam-tree-won-viewers-hearts-247463

Bunyan, J. (2020). Deputy minister who accused Sabah student Veveonah of lying says info came from Kudat MP. *Malay Mail.* https://www.malaymail.com/news/malaysia/2020/09/07/deputy-minister-who-accused-sabah-student-veveonah-of-lying-says-info-came/1900936

Chen, Y.C. (2017). Digital divide, digital inclusion, and digital opportunities. *Managing Digital Governance*, 54-75. doi:10.4324/9781315207667-10

Chung, H. L. (2023). Sarawak to launch blueprint to becoming leading digital economy by 2030. *The Edge Malaysia*. https://theedgemalaysia.com/article/sarawak-launch-blueprint-becoming-leading-digital-economy-2030

National Water Research Institute. (2021). *Climate change adaptation framework for water sectors*. National water research institute of malaysia ministry of environment and water.

Dawood, S. R. (2019). Digital divide and poverty eradication in the rural region of northern peninsular Malaysia. *The Indonesian Journal of Geography, 51*(2), 172. doi:10.22146/ijg.37758

Desormeau, P. A., Walsh, K. M., & Segal, Z. V. (2018). Mindfulness-based stress reduction and mindfulness-based cognitive therapy. The Oxford Handbook of Stress and Mental Health, 688-704. Oxford Press. doi:10.1093/oxfordhb/9780190681777.013.33

Dijk, J. V. (2020). *The digital divide*. John Wiley & Sons.

Doksil, M. (2020, June 24). UMS student who climbed hill, tree for better internet connectivity gets aid. *The Borneo Post.* https://www.theborneopost.com/2020/06/24/ums-student-who-climbed-hill-tree-for-better-internet-connectivity-gets-aid/

Economic Planning Unit. (2022). NATIONAL [Economic Planning Unit Prime Minister's Department.]. *Energy Policy*, 2022–2040.

Editoron. (2021, December 14). Budget 2022 to boost East Malaysia's development. *The Borneo Post.* https://www.theborneopost.com/2021/12/14/budget-2022-to-boost-East-malaysias-development/

Ferreira, D., Vale, M., Miguel Carmo, R., Encalada-Abarca, L., & Marcolin, C. (2021). The three levels of the urban digital divide: Bridging issues of coverage, usage and its outcomes in VGI platforms. *Geoforum, 124*, 195–206. doi:10.1016/j.geoforum.2021.05.002

Fujita, S. (2021). *Income inequality in terms of a Gini coefficient: A Kaleckian perspective.*

Gonzales, A. (2015). The contemporary US digital divide: From initial access to technology maintenance. *Information Communication and Society, 19*(2), 234–248. doi:10.1080/1369118X.2015.1050438

Guo, C., & Wan, B. (2022). The digital divide in online education in China during the COVID-19 pandemic. SSRN Electronic Journal. doi:10.2139/ssrn.4068301

Hanafizadeh, M. R., Hanafizadeh, P., & Bohlin, E. (2013). Digital divide and E-Readiness. *International Journal of E-Adoption, 5*(3), 30–75. doi:10.4018/ijea.2013070103

Hanafizadeh, M. R., Hanafizadeh, P., & Saghaei, A. (2013). Perusing E-readiness and digital divide. *Digital Literacy*, 313-346. doi:10.4018/978-1-4666-1852-7.ch017

Heimann, M., Bus, A., & Barr, R. (2021). Growing up in a digital world - Social and cognitive implications. Frontiers Media SA.

Jacobs, J., & Subramaniam, P. (2020, October 23). Cover story: The digital divide and disconnection. *The Edge Markets.* https://www.theedgemarkets.com/article/cover-story-digital-divide-and-disconnection

Jamil, S. (2021). From digital divide to digital inclusion: Challenges for wide-ranging digitalization in Pakistan. *Telecommunications Policy, 45*(8), 102206. doi:10.1016/j.telpol.2021.102206

Johari, Z. (2020, July 19). Grouses over poor internet access. *The Boreno Post.* https://www.theborneopost.com/2020/07/19/grouses-over-poor-internet-access/

Kementerian Komunikasi Dan Multimedia Malaysia. (2018). *Communications & multimedia blueprint 2018-2025 digitise and humanise* (1st ed.). Ministry of Communications and Multimedia.

Klaus, M. (2017). *Digital and the real world, the: Computational foundations of mathematics, science, technology, and philosophy*. World Scientific.

Kreitem, H. M. (2020). Weaponization of access, communication inequalities as a form of control: Case of Israel/Palestine. *Digital Inequalities in the Global South*, 137-157. doi:10.1007/978-3-030-32706-4_7

Kuráková, I., Vallušová, A., & Marasová, J. (2021). Measuring the digital divide in the V4 countries using the digital divide index. *Ekonomika a spoločnosť, 22*(1), 77-93. doi:10.24040/eas.2021.22.1.77-93

Lahtinen, O., & Salmivalli, C. (2020). An effectiveness study of a digital mindfulness-based program for upper secondary education students. *Mindfulness, 11*(11), 2494–2505. doi:10.100712671-020-01462-y

Malay Mail. (2023, March 17). *Sabah state govt completes three digital hubs.* Malay Mail. https://www.malaymail.com/news/malaysia/2023/03/17/sabah-state-govt-completes-three-digital-hubs/60050

Malaysian communications and multimedia commission. (2023). *Capacity Development Programme (CDeP) 2023 The Evolution of The Mobile Cellular Communication: From 0G to 5G.* https://www.mcmc.gov.my/en/capacity-development-programme-cdep/capacity-development-programme-cdep-2023

MIGHT. (2014). *Malaysian Rail Supporting Industry Roadmap 2030.* Malaysian Industry-Government Group for High Technology.

Minister of Finance. (2022). *Budget Speech 2023.* MOF. chrome-extension://efaidnbmnnnibpcajpcglclefindmkaj/https://budget.mof.gov.my/pdf/2023/ucapan/buku-budget-speech-2023.pdf

Muhammad, A. (2022, April 8). 'I travel 10-15km daily just to get internet': Telupid student attends online class in car boot. *Daily express online.* https://www.dailyexpress.com.my/news/190389/i-travel-10-15km-daily-just-to-get-internet-telupid-student-attends-online-class-in-car-boot-/

Mukhopadhyay, N., & Sengupta, P. P. (2021). *Gini inequality index: Methods and applications.* CRC Press. doi:10.1201/9781003143642

OECD. (2001). *Understanding the digital divide.* OECD Digital Economy Papers. doi:10.1787/20716826

One, T. M. (2023, February 8). *Pushing Malaysia's smart city development in 2022.* TMONE. https://www.tmone.com.my/resources/think-tank/article/smart-city-development-in-2022/

Piaget, J. (1964). Part I: Cognitive development in children: Piaget development and learning. *Journal of Research in Science Teaching, 2*(3), 176–186. doi:10.1002/tea.3660020306

Prime Minister Malaysia. (2015). *Eleventh Malaysia plan, 2016-2020: Anchoring growth on people.* Unit Perancang Ekonomi.

Reibel, D., & McCown, D. (2019). Mindfulness-based stress reduction. Handbook of Mindfulness-Based Programmes, 29-41. Oxford Press. doi:10.4324/9781315265438-4

Scheerder, A., Van Deursen, A., & Van Dijk, J. (2017). Determinants of internet skills, uses and outcomes. A systematic review of the second- and third-level digital divide. *Telematics and Informatics*, *34*(8), 1607–1624. doi:10.1016/j.tele.2017.07.007

Schenkman, L. (2021). Brain's sensory processor may prompt memory problems in autism. *Spectrum (Lexington, Ky.)*. doi:10.53053/INWE5797

SEDA. (2023a). *LSS – Renewable energy Malaysia*. SEDA Malaysia. https://www.seda.gov.my/reportal/large-scale-solar/

SEDA. (2023b). *Net Energy Metering (NEM) program*. SEDA.

Shamsul Bahar, M. N. (2023). *CEO's message – Malaysian green technology and climate change Corporation*. Malaysian Green Technology And Climate Change Corporation – Leading Partner for Green Growth & Climate Change. https://www.mgtc.gov.my/who-we-are/ceos-message/

Surianshah, S. (2021). Digital divide in education during COVID-19 pandemic. *Jurnal Ekonomi Malaysia*, *55*(3), 103–112. doi:10.17576/JEM-2021-5503-07

Svongoro, P. (2022). *The digital divide in the context of E-learning during the COVID-19 pandemic: A zimbabwean perspective*. doi:10.25159/UnisaRxiv/000041.v1

The Malaysian Industry-Government Group for High Technology (MIGHT). (2023). *MIGHT Portal*. Malaysian Industry-Government Group for High Technology. https://might.org.my/

Valk, R., & Planojevic, G. (2021). Addressing the knowledge divide: Digital knowledge sharing and social learning of geographically dispersed employees during the COVID-19 pandemic. *Journal of Global Mobility*, *9*(4), 591–621. doi:10.1108/JGM-02-2021-0019

Van de WerfhorstH. G.KessenichE.GevenS. (2020). The digital divide in online education. Inequality in digital preparedness of students and schools before the start of the COVID-19 pandemic. doi:10.31235/osf.io/58d6posf.io/58d6p

Van Deursen, A. J., & Helsper, E. J. (2015). The third-level digital divide: Who benefits most from being online? *Communication and Information Technologies Annual*, 29-52. doi:10.1108/S2050-206020150000010002

Van Deursen, A. J., & Van Dijk, J. A. (2018). The first-level digital divide shifts from inequalities in physical access to inequalities in material access. *New Media & Society*, *21*(2), 354–375. doi:10.1177/1461444818797082 PMID:30886536

Van Deursen, A. J., & Van Dijk, J. A. (2019). The first-level digital divide shifts from inequalities in physical access to inequalities in material access. *New Media & Society*, *21*(2), 354–375. doi:10.1177/1461444818797082 PMID:30886536

Veveonah, M. (2020, June 16). *24 Hours on tree [Video].* YouTube. https://www.youtube.com/watch?v=-qxTyBWVLIU

Vinod, G. (2021). Budget 2022: Increase allocation by RM2 bil to help East malaysians, non-malays. *Focus Malaysia.* https://focusmalaysia.my/budget-2022-increase-allocation-by-rm-2-bil-to-help-East-malaysians-non-malays/

Yahaya, N. (2020, September 10). KJ jumpa Veveonah, minta maaf wakili kerajaan. *Sinarharian.* https://www.sinarharian.com.my/article/100550/berita/nasional/kj-jumpa-veveonah-minta-maaf-wakili-kerajaan

Yap, M. Y. (2020, June 15). This Malaysian student spent 24 hours on a tree in order to get internet signal to complete her exams. *Mashable.* https://sea.mashable.com/culture/11036/this-malaysian-student-spent-24-hours-on-a-tree-in-order-to-get-wi-fi-signal-to-complete-her-exams

Chapter 2
Economic Benefits of High-Speed Broadband Coverage and Adoption

Wolfgang Briglauer
EcoAustria, Austria

Nicole Palan
University of Graz, Austria

ABSTRACT

Public investment into the availability of ICT infrastructure has been studied for decades to assess its effect on economic development and productivity growth. Whereas in the 1980s, this referred to the effects of telephones, since the 1990s it is the availability of internet connections, which is of great interest for politicians and policy makers alike. During the Covid-19 pandemic, social distancing has led to a transformation of work relations and the widespread introduction of work from home. Also, distance education as well as the increased importance of online streaming portals for leisure activities proved the necessity for fast broadband coverage and adoption. The authors investigate the degree of broadband coverage with a special focus on Austria and provide a survey on the economic benefits especially of high-speed internet connections for the economic development. The authors show to which extent Austria is using the full potential of fast and ultrafast internet connections and highlight the benefits for both firms and inhabitants in both rural and urban environments.

DOI: 10.4018/978-1-6684-8634-4.ch002

INTRODUCTION

Economic policy makers and empirical researchers alike try to answer how important the availability and adoption of (fast) broadband internet has been for the economic development of regions and entire countries. In this context, the European Commission (2022a) has highlighted the importance of economic policy measures to "*enable access to digital services for all citizens and to maintain its prosperity*". As one of the most important economic policy agendas of the European Union is cohesion, the full-coverage of broadband internet for ALL citizens – irrespective of their socio-economic background and location – is essential. To promote competitiveness and economic development, "*the European Union needs a digital connectivity infrastructure of top performance, security and sustainability, optimised to leverage the latest optical fibre technologies in fixed networks and to connect innovative wireless systems such as 5G, 6G and Wi-Fi.*"

Therefore, the European Union as well as the individual member countries have set ambitious targets to implement and upgrade the digital infrastructure. For example, the European Union launched a "5G Action Plan" (European Commission, 2016) or the "Gigabit Strategy" and the Digital Agenda Europe objectives (European Commission, 2010). The main objectives that should be achieved by 2025 are the following: a) Gigabit connectivity for all main socioeconomic drivers, b) uninterrupted 5G-coverage for all urban areas and main transportation paths and c) access to at least 100 Mbps broadband internet for all European households. By 2030 the goal is to cover all European households by a Gigabit network and to cover all populated areas within the European Union are with 5G.

The Covid 19 pandemic made the need for a well-developed, ubiquitous, and stable digital infrastructure to sustain business activities, education, and social relations even more evident. Due to the imposition of social distancing measures worldwide, major parts of the labor force for instance experienced teleworking for the very first time during their careers. In the case of Austria, more than two third of all home office workers in 2020, had never worked from home before. The pandemic and the ongoing trend for home office led to a substantial re-organization of work, leading also to changes in the demand for (fast) internet connections by many households as an appropriate infrastructure is needed for remote work (Bachmayer and Klotz, 2021). In this respect, the stable availability of fast internet connections has become a key factor for many employees: Firms allow remote work if employees can work as effectively from home as in the office, implying the need for the frictionless access to firm data and the ability to easily join online meetings.

The Covid-19 pandemic also showed the problems associated with a lack of digital infrastructure and the need to incorporate digital competences in the field of education. Digital connectivity proved to be vital to access teaching units, get feedback from teachers, interact with classmates, and share information. Digital tools enabled millions of pupils and students to learn from home, access books and learning material but also to have entertainment during leisure when theaters, cinemas, museums, and operas had to close. In this context, not only the access to digital infrastructure but also the digital skills needed to use teaching material, take part in online meetings proved to be critical aspects, as not all households have equal access and equal skills to use digital infrastructure and digital tools (van de Werfhorst et al. 2022).

BACKGROUND

Information and communication technologies (ICT) in general – and broadband internet more specifically – have shown their power in transforming communication, the organization of work, the way business is conducted and knowledge transferable. As these changes affect many sectors in the economy and fundamentally alter the socio-economic organization of work, this process is also referred to as the "*ICT revolution*" or "*Third Industrial Revolution*" (Rifkin, 2013). Bresnahan and Trajtenberg (1995) refer to those technologies as general-purpose technologies, when they allow for new methods in production and this effect not being limited to a small number of ICT producing sectors. Thus, these new production methods and inventions have the chance to alter existing social and economic structures to a great extent. Jovanovic and Rousseau (2005) as well as Basu and Fernald (2007) identified the Internet as one of the two most important and the latest GPT respectively. The transformational impact of information and communication technologies on the socio-economic environment of advanced economies is evident, the link between ICT in general and (fast) broadband internet more specifically is clear. Especially at the beginning of the implementation of general-purpose technologies, the impact on productivity and economic growth can be ambiguous or even negative, as learning costs are high and large investments are necessary to benefit from these GPTs as stressed by Jovanovic and Rousseau (2005). Moreover, it should be kept in mind that broadband applications can also improve the efficiency of labor and thereby act as a substitute for labor, increasing productivity and economic growth while reducing employment. In general, the effect of broadband internet on employment depends also on population trends as well as the overall state of the economy. In times of full employment, the adoption of broadband connections is unlikely to increase employment as much as during an economic recession, when there are many unemployed people searching for a job. Irrespective of the business cycle, the access and use of broadband applications can however reduce frictional unemployment due to better and faster access to information but lead to structural unemployment due to changes in digital skills required for the job market (Holt and Jamison, 2009).

Taking a closer look at the availability of broadband internet, the fight against the "*digital divide*" became an important policy goal. The "digital divide" describes problems associated with the unequal opportunities in accessing and using information and communication technologies by individuals and firms based on their geographical location, income and education levels and being the member of a minority (OECD, 2001). In the 1990s and 2000s the digital divide referred to the differences in access rates to first-generation broadband networks, dividing communities and countries into those who had access and those who had no access at all (Ragnedda and Muschert, 2013). The primary policy goal during these decades was to provide internet to *all* municipalities of a country and thus foster the deployment in rural areas to catch up with urban regions. At that time, there were high hopes that the digital divide could soon be closed, especially with the help of government subsidies.

Yet, at that time politicians and policy advisers underestimated the development concerning the quality improvements of broadband infrastructure and the importance upgrading the broadband infrastructure. Therefore, the deployment of high-speed networks has become a major challenge for regulators, governments, and network providers: On the supply side, high-speed broadband coverage is an even more intensive and expensive investment than basic broadband internet due to high construction costs related to digging and laying down fiber-optical cables compared to first generation broadband infrastructure. As private companies seek to maximize profits, it is highly unlikely that private investors offer fast broadband infrastructure in areas with low population densities. This is because costs of deployment can only be divided between a low number of firms and households, leading to higher average deployment costs in rural areas compared to urban regions. Consequently, any ubiquitous coverage targets are likely to require additional funding from governments.

A further challenge is to be found on the demand side: Evidence shows that already deployed fiber-based broadband infrastructures are still substantially under-utilized in many countries (European Commission, 2022a and RTR, 2022 for Austria in particular). In particular, the average consumer adoption rates in most developed countries are still far below coverage levels. Finally, on the demand side a further point of concern attracted attention in the digital divide policy: Improving the digital skills of consumers and employees are becoming more and more important to make use of the full potential of broadband infrastructures and their impact on economic growth and productivity.

MAIN FOCUS OF THE CHAPTER

The central aim of this book chapter is to provide evidence on the correlation between the availability of (fast) broadband internet connections and economic growth, productivity and (un)employment. Moreover, this chapter provides a special focus on the status quo of the available digital infrastructure in Austria. To this end, the chapter discusses the achievements and challenges in providing ubiquitous (fast) internet in all regions of Austria.

Bertschek et al. (2016) provided an important review of more than sixty studies on the causal effects of general broadband coverage and adoption on economic variables such as GDP growth, employment, and productivity. Many of these studies confirmed a positive impact of broadband adoption on economic growth. Similarly, most studies found a positive link between the availability of broadband and employment growth. However, there are notworthy differences between the studies regarding the impact of digital infrastructure and this comparative survey also did not focus on the effects of fast internet connections.

Qiang and Rossotto reported in a World Bank study in 2009 that a ten percent increase in broadband penetration increses economic groth by 1.2 percentage points in developed countries and by almost 1.4 percentage points in low and middle-income countries. According to Greenstein and McDevitt (2011) the contribution of broadband infrastructure on US-American GDP in 2006 was 28 billion Dollar. Moreover, broadband accounted for 8.3 to 10.6 billion of additional revenue. Forman et al. (2012) could show that in those US counties with high population density, high IT-intensity and a higher proportion of inhabitants with high incomes and higher education, the advanced internet investments could explain more than a fifth of the wage growth. In contrast, in regions with more unfavorable conditions only one per cent of wage growth could be related to advanced internet investments. Thus, both the geographical and the socio-economic background of inhabitants play an important role whether the supply of broadband infrastructure can materialize the potential benefits. This means in particular that the demand for high broadband internet might vary between regions and the positive effects of supply can only materialize if a) sufficient people are highly educated to be in need for fast broadband internet and b) enough IT-intensive firms are located in the respective region.

Investigating Norwegian municipalities, Akerman et al. (2015) reported that an increase in broadband availability of one per cent led to an increase in the wages of skilled workers by 0.013 per cent. Similarly, regions where more high-income inhabitants lived who used IT more intensively, there was a stronger impact from broadband infrastructures and service adoption on wages. This effect was even larger for remote areas than for urban areas. Likewise did Atasoy (2013) report positive effects of broadband availability on the employment rate at the US county level for the period 1999 to 2007. On the regional level, rural areas benefited more from increased broadband availability than urban areas. In the extreme case of moving from no broadband availability to full broadband availability, the positive effect on the employment rate would have been an increase of 2.24 percentage points. Studying more than 4300 Canadian communities, Ivus and Boland (2015) found no significant impact of fixed broadband coverage on local employment and wage growth for the period of 1997 to 2011 when studying the isolated effect of an increase in broadband coverage. However, when an interaction term between broadband coverage and rural areas was included, the authors found a significant positive impact of higher broadband coverage on employment growth in rural areas, while they report a significant negative impact effect on employment growth in urban areas. In this study, these effects also varied across industries, with particularly strong effects in IT-intensive industries. This relatively larger impact of broadband in rural areas was also supported by by Kolko (2012) and by Whitacre et al. (2014a, 2014b). Focusing on the impact of broadband infrastructure on employment growth, Gillett et al. (2006) as well as Crandall et al. (2007) show a positive effect at the regional level. The results imply that if broadband capacity per capita increases, the number of jobs in these regions also increased.

Regarding fiber-based broadband, the results are more heterogenous: On the one hand, Katz and Callorda (2020) found that an increase in fiber-based broadband speed by 100 per cent increases the overall employment in all sectors by around .23 per cent, and in the service sector by around 1.53 per cent. Also Briglauer et al. (2021) analyzed German county-level data and reported substantial positive spillover effects among neighbouring counties, with more pronounced effects for rural regions. Katz and Callorda (2020) reported for US data that GDP increased by .73 per cent if fiber-based cable broadband speed accelerated by 100 per cent. On the other hand, earlier studies by Bai (2017) (with corrections identified by Whitacre et al. (2018)) and Fabling and Grimes (2021) could not identify a significant impact of fiber-based broadband on employment. Nordin et al. (2019) even presented negative results for the whole sample and only determined a positive effect when considering rural areas. Balsmeier and Woerter (2019) only obtained a positive impact of fiber-based broadband availability on high-skilled employment in Switzerland, yet no no statistically significant and no positive results for the entire sample (i.e. not splitting the sample into high-, middle- and low-skilled workers).

In recent years, an additional focus of the benefits of broadband infrastructure has been added: Studies now also add to our understanding of increases in the consumer surplus due to advances in broadband infrastructure, thereby shifting the point of analysis towards the benefits of broadbant internet that stem from internet adoption by consumers. The first estimates of Greenstein and McDevitt (2011) showed that the consumer surplus ranged between 4.8 and 6.7 billion dollar and that households generated 20 to 22 billion Dollar revenue from broadband. Dutz et al. (2012) compared the availability of basic broadband at home compared to no broadband access and estimated the additional consumer surplus benefits of broadband access in 2008 to be around $32 billion. More recent studies, like Katz et al. (2020) found that between 2021 and 2027, the tremendous rise in broadband speed will increase the average annual consumer surplus by $71.5 billion.

The Status Quo of Broadband Availability in Austria

The European Commission regularly presents the Digital Economy and Society Index (DESI) which analyzes the development of EU member countries regarding a variety of indicators related to digitalization. The results for 2021 showed that Austria has an above EU-average coverage in 5G, yet "*continues to perform significantly below the EU average for fixed very high-capacity network coverage and take-up. Ensuring access to higher speeds in rural areas remains Austria's biggest hurdle to achieve nationwide Gigabit connectivity by 2030.*" (European Commission, 2022b). Also, Bärenthaler-Sieber et al. (2022) and RTR (2022) highlight that Austria is lagging especially with respect to the adoption of ultrafast internet connections even though contract costs are comparatively low for consumers. There are two main factors that could explain this situation due to supply-side reasons: First, the topography with many mountain areas makes it difficult and cost-intensive for firms to deploy broadband infrastructure. Second, people in Austria live rather dispersed compared to other countries. This low population density is a disadvantage as it further drives up costs of building broadband infrastructure. Looking at the demand side, RTR (2022) reported that due to the need for the smooth functioning of video conferences in home office, the increased use of streaming services and the use of smart home solutions, the demand for high-bit-rate broadband connections has increased steadily. Data showed that the data volume usage in both fixed and mobile networks had been rising significantly over the years with mobile communication devices continuously playing an essential role in the broadband supply of households in Austria. For example, the number of connections supporting bandwidths of more than 100 Mbit/s increased by almost thirty per cent within one year, accounting for more than a quarter of all fixed broadband connections in 2021.

Regarding the overall coverage with fixed broadband connections in all Austrian federal provinces, remarkable differences between the nine provinces persisted. Whereas there were provinces like Salzburg and Vienna with almost entire broadband penetration, the provinces of Carinthia, Lower and Upper Austria still lagged. Especially, in rural areas, penetration with bandwidths above 30 Mbit/s or even 100 Mbit/s remained low. Gradual improvements did take place, yet it remained obvious that there is a continued high need for action to promote the availability and use of fiber-based digital infrastructure in Austria (RTR, 2002 and European Commission, 2022b). Comparing with other EU countries, Austria is above the EU average regarding 5 G coverage and adoption but is below the EU average when it comes to fixed very high-capacity broadband infrastructure. The DESI report 2022 points out that "*Ensuring access to higher speeds in rural areas remains Austria's biggest hurdle to achieve nationwide Gigabit connectivity by 2030.*" (European Commission, 2022b).

Regarding the financial aspects of the broadband roll-out, Austria has opted for a market-driven network approach, which means that the Austrian governments aims to limit the use of public funds only to those provinces with the lowest chances of benefiting from private sector investments. Thus, government intervention should be limited mainly to disadvantageous peripheral location (Federal Ministry for Transport, Innovation and Technology, 2019). As the Austrian government is aware of the economic and social benefits of digitalization and aims to fully exploit this potential for the entire population and all companies, the ubiquitous availability of gigabit-capable broadband infrastructure is of essential importance as outlined by the Austrian "Broadband Strategy 2030".

According to Statistics Austria, about 59 per cent of Austrian inhabitants lived in urban regions and cities with more than 10.000 inhabitants in 2021. Even though the degree of urbanization of the Austrian population has been rising over the years, still almost forty per cent of the population live in rural areas (Statistics Austria, 2021). Therefore, it is of great importance that Austria has achieved significant improvements regarding the access to very fast broadband in rural areas. In 2020, the coverage of fast broadband in rural areas was less than forty percent and could be increased to almost 70 per cent within one year. Yet, when it comes to fixed very high-capacity network coverage, Austria is lagging even further behind compared with other European countries. Despite the increase in the coverage of the entire country from 14 to 39 percent from 2019 to 2020, further increases in 2021 were rather modest leading to a coverage in Austria of 45 per cent at the end of 2021. Focusing only on rural areas in Austria, the situation is even worse: The coverage in these areas only increased by 6 per centage points from 2019 to 2021, thus the gap between urban centers and rural areas increased. Moreover, when compared with the EU-average, the development in Austrian rural areas is far less favorable: Whereas the coverage of ultrafast internet connections in rural areas was fifty per cent on the EU-average, the coverage in Austria was only 27 per cent (European Commission, 2022b). As pointed out before, the problem in Austria is also to be found on the demand side: Incentives for firms to pursue the expansion of ultra-fast broadband (i.e. with download rates of at least 100 Mbit/s) are lower than in other European countries as the demand for high bandwidth internet in Austria is much less dynamic, making market penetration for fast internet connections less profitable. As pointed out before, often the existing capacities are not being utilized enough by consumers even though the costs for these fast internet connections are considerably low. In this context, Bärenthaler-Sieber et al. (2018) spoke of a demand gap that increases with rising download rates for Austria.

SOLUTIONS AND RECOMMENDATIONS

Broadband and the Labor Market

According to Eurostat (2022), on average twelve per cent of employed people in the EU regularly worked from home in 2020, the first year of the Covid-19 pandemic. Estimates from September 2022 show a further increase in 2021 by 1.4 per centage points. This highlights the ongoing transformation processes in the organization of work within companies and the trend towards continuous remote work beyond the immediate needs during lockdowns and regulations about social distancing. Remarkable differences between European countries exist, however: Whereas in Ireland one out of three employees regularly worked from home in 2021, only three out of 100 Bulgarians did so. These differences in “teleworkability” (Brussevich et al. 2022) do not only highlight discrepancies regarding the education level, the skill and employment structure of the workforce (in this respect, the share of white-collar employment is essential), work culture and legal requirements, but can also partly be explained by the heterogeneity in the availability of (fast) broadband internet and thus the technical pre-requisites for work from home. Brussevich et al. (2022) show that additionally to the availability of broadband at home, business’ internet connectivity is of equal importance, which is measured by the average percentage of employees who regularly use a computer with internet access in their job. In this respect, the study identified the lack of broadband infrastructure due to too little investment into information and communication technologies in general as problems. Often, the country differences in costs for companies to implement fast broadband connections explain the differences in firm uptake of the latest digital infrastructure available.

Government subsidies providing fast broadband internet to remote areas, firms and households can thus both increase the share of employees who can benefit from remote work solution and make firms in peripheral regions more competitive. In times when the European Union and many countries worldwide seek to reduce CO_2-emissions, investments in broadband network can also positively affect the total amount of emissions as commuting traffic is reduced and can add to the transformation towards a greener society (European Commission, 2021). Firms thus should use the possibilities to further reduce and partially substitute business travels when beneficial by online meetings and video conferences. Governments should facilitate and could allow for (further) tax benefits for remote working solutions. Also, students benefit from less commuting if online study programs help them minimize travel time. In this respect, (partly) online study programs can become of greater importance also to allow for lifelong learning and skill upgrade of the existing workforce. In this respect, the skills needed to use online learning platforms are essential and it could be beneficial if governments provide and subsidize programs to acquire these digital skills to keep the workforce up with the ever-increasing need of digital skill demand.

Broadband Investments

The public funding for the expansion of fiber-based high-bandwidth networks can be justified from a distributional point of view to combat the digital divide and the consequences thereof. Empirical evidence has shown that there are high risks of underinvestment in peripheral regions with low population density if governments do not provide funding as private companies often do not invest enough due to a cost-benefit analysis (which does however not consider the positive externalities of this investment). Austria is a case in point: In both rural areas and mountain regions with low population density, investment in fiber-based broadband networks hardly pays off for firms and therefore private companies do not undergo deployment. Yet as shown before, high-skilled people living in these regions would benefit the most from the enhanced availability of the fast broadband infrastructure, decreasing the likelihood of these highly skilled people from moving to urban areas. Thereby the convergence of regions and decreasing the push-effects of peripheral regions could be accomplished.

As pointed out before, population density has considerable consequences for the deployment costs of broadband internet. At the same time, regional governments of peripheral regions more and more highlight the need for fast internet connections for the future economic development and competitiveness. Politicians in more remote areas thus put high hopes on the fact that the availability of fast broadband infrastructure can facilitate decentralized work in peripheries and decrease migration from these areas to urban centers. Empirical evidence by Briglauer et al. (2018) showed that in addition to deployment costs, regulation in broadband markets decreases the incentives of private firms to invest. De-regulation and the removal of bureaucratic procedures could promote investments and facilitate the deployment of broadband infrastructure.

Investments into faster broadband internet and a wider availability thereof can also be justified on the grounds of positive externalities. Broadband internet can lead to cost-reducing effects for a wide range of sectors in the economy and to additional economic growth effects in rural regions.

Due to the different roll-out costs for the different broadband technologies and due to substantial uncertainties about technological advances and the further development of demand, it is beneficial if public funding models are technology-neutral and restrictive regarding the targeted funding areas (Briglauer et al., 2020). In addition to the efficiency of funding measures, however, it is also important to consider their effectiveness: A case in point are existing federal and state level funding channels, which demonstrate that a simplification of sometimes complicated and bureaucratic application procedures would be necessary (Bertschek and Briglauer, 2018).

Regarding financial effectiveness, it is important for governments to concentrate the provision of digital infrastructure in those areas that would not be covered by private providers (i.e. regions in which high deployment costs outweigh the ability to pay of private consumers). Therefore, it is essential that policy makers well understand to which extent private providers provide broadband infrastructure and to which extent the government needs to act as a provider to avoid too little coverage. In this process, it is also necessary to co-ordinate public with private investment to achieve the goal of ubiquitous availability of the latest broadband infrastructure available.

CONCLUSION

This chapter gave an overview of the existing literature on the positive impact of broadband infrastructure on employment, productivity, and economic growth. Besides the benefits for process innovation and productivity, (fast) broadband infrastructure can also have a significantly positive impact on consumers and employees. These benefits proved to be especially important during lockdowns in 2020 and 2021. Regarding labor market effects, broadband internet has facilitated work from home and continues to be important as work environments and the preferences of employees change. Yet, to use the full potential of home office also in remote areas, stable and often fast internet connections are a pre-requisite to take part in online conferences, etc. Thus, advances in the availability of fast broadband internet connections are expected to generate new/more employment opportunities especially in rather remote areas, as it enables employees to work from home, reduces the need to live near work and reduces commuting time.

The DESI 2022 report (European Commission 2022a) also made clear that businesses only slowly adopt digital technologies such as artificial intelligence and big data. As digital skills (both basic and advanced) are important for future prospects of economic development, it is of great economic policy importance to prevent digital exclusion and to close the digital divide not only on the provision of fast and ultrafast broadband internet connections but also regarding the adoption of this infrastructure.

Due to the high relevance of (fast) broadband connections for the social and economic wellbeing of households and businesses in a country, governments should therefore pursue and help finance the full coverage of this infrastructure. Therefore, public support measures to finance the expansion of high-broadband networks is crucial and needs to be fostered even in times of rising public debt levels when other pressing economic challenges such as high inflation rates in 2022. In this context it is noteworthy that the literature provides ample evidence that the full realization of broadband targets for all households and firms and the associated public funding needed can be justified from a distributional point of view as the universal access to fast broadband infrastructure can reduce both the social and spatial digital divide. Recent years have shown that it is also important to acknowledge that the availability of broadband internet does not automatically lead to the adoption of broadband services by consumers and firms. Hence, it is necessary to influence the factors that facilitate the full usage of broadband supplied such as awareness of the availability, the price of latest infrastructure as well as digital literacy such that people can make use of the benefits of fast internet connections. In this context, the OECD (2018) highlights the role governments can play in raising awareness and increasing the digital literacy of its inhabitants to encourage so far non-users to engage in the use of broadband-based services and increase the e-literacy of the population. To that end, governments can set measures to increase the ICT affinity of its population and for instance install the compulsory subject "digital competences" at schools. In Austria, for instance, the subject "Digital Literacy" has been introduced at secondary schools in the school year 2022/2023 and offers one hour per week from grade five to eight (European Commission, 2022b). The Austrian Ministry of Education, Science and Research moreover launched a detailed plan for "Digital Education", according to which all federal schools should have fiber-optic broadband connections and a high-performance WLAN supply in Austrian classrooms by 2023. To promote the expansion of fiber-based networks in compulsory schools, initial connections are publicly funded by 90 per cent with a total funding of 28.8 million Euros. Moreover, this plan aims for the equipment of a digital end device (either a tablet or a computer) for all pupils from 5th to 8th grade between 2021 and 2026.

To sum up, the lack of access to broadband internet in general and fast broadband internet in particular, can create and/or amplify significant economic as well as social problems in various dimensions: As (high-speed) broadband networks are likely to be key technologies, which do not only influence productivity, employment and economic growth in sectors directly related to IT, a lack of this infrastructure is likely to have a negative impact on economic growth, on labor productivity, on employment creation in regions with poor broadband networks. Moreover, a negative impact on the promotion of innovation and international collaborations as well as a limited ability to take part in education and recreation activities is likely. Also, the lack of access to fast broadband infrastructure means that firms which more and more rely on e-commerce are diverted from regions with unfavorable conditions regarding the availability of internet connectivity. Governments should therefore address the challenges associated with the digital divide and improve the universal access to technological infrastructure and the use thereof.

ACKNOWLEDGMENT

The authors are grateful to having received funding from the Oesterreichische Nationalbank for for the research project Jubiläumsfonds Nr. 18736.

REFERENCES

Akerman, A., Gaarder, I., & Mogstad, M. (2015). The Skill Complementarity of Broadband Internet. *The Quarterly Journal of Economics*, *30*(4), 1781–1824. doi:10.1093/qje/qjv028

Atasoy, H. (2013). The effects of broadband internet expansion on labor market outcomes. *Industrial & Labor Relations Review*, *66*(2), 315–345. doi:10.1177/001979391306600202

Bachmayer, W., & Klotz, J. (2021). *Homeoffice: Verbreitung*. Gestaltung, Meinungsbildung und Zukunft.

Bai, Y. (2017). The Faster, the Better? The Impact of Internet Speed on Employment. *Information Economics and Policy*, *40*, 21–25. doi:10.1016/j.infoecopol.2017.06.004

Balsmeier, B., & Wörter, M. (2019). Is this time different? How digitalization influences job creation and destruction. *Research Policy*, *48*(8), 62–73. doi:10.1016/j.respol.2019.03.010

Bärenthaler-Sieber, S.; Bock-Schappelwein, J.; Böheim, M.; Kügler, A. and N. Schmidt-Padickakudy (2022). Digitalisation in Austria: Progress, Broadband Infrastructure and Open Access Networks. *WIFO Reports on Austria*. WIFO.

Basu, S., & Fernald, J. (2007). Information and Communications Technology as a General-Purpose Technology: Evidence from US Industry Data. *German Economic Review*, *8*(2), 146–173. doi:10.1111/j.1468-0475.2007.00402.x

Bertschek, I. & Briglauer, W. (2018). *Wie die Digitale Transformation der Wirtschaft gelingt*. ZEW-Policy Brief 5/2018.

Bock-Schappelwein, J., Firgo, M., Kügler, A., & Schmidt-Padickakudy, N. (2021). *Digitalisierung in Österreich: Fortschritt, digitale Skills und Infrastrukturausstattung in Zeiten von Covid-19*. WIFO Monatsberichte 6/2021. https://www.wifo.ac.at/jart/prj3/wifo/resources/person_dokument/person_dokument.jart?publikationsid=67254&mime_type=application/pdfwifo.ac.at)

Bresnahan, T., & Trajtenberg, M. (1995). General Purpose Technologies 'Engines of Growth'? *Journal of Econometrics*, *65*(1), 83–108. doi:10.1016/0304-4076(94)01598-T

Briglauer, W., Cambini, C., & Grajek, M. (2018). Speeding Up the Internet: Regulation and Investment in the European Fiber Optic Infrastructure. *International Journal of Industrial Organization*, *61*(C), 613–652. doi:10.1016/j.ijindorg.2018.01.006

Briglauer, W., Dürr, N., & Gugler, K. (2021). A retrospective study on the regional benefits and spillover effects of high-speed broadband networks: Evidence from German counties. *International Journal of Industrial Organization*, *74*, 102677. doi:10.1016/j.ijindorg.2020.102677

Briglauer, W., & Gugler, K. (2019). Go for Gigabit? First Evidence on Economic Benefits of High-Speed Broadband Technologies in Europe. *Journal of Common Market Studies*, *57*(5), 1071–1090. doi:10.1111/jcms.12872

Briglauer, W. and V. Stocker (2020). Bedeutung digitaler Infrastrukturen und Dienste und Maßnahmen zur Förderung der Resilienz in Krisenzeiten. *EcoAustria Policy Note 42*.

Briglauer, W., Whalley, J., & Stocker, V. (2020). *Public policy targets in EU broadband markets: The role of technology neutrality* (Vol. 44). Telecommunication Policy.

Brussevich, M., Dabla-Norris, E., & Khalid, S. (2022). Who bears the brunt of lockdown policies? Evidence from teleworkability measures across countries. *IMF Economic Review*, *70*(3), 560–589. doi:10.105741308-022-00165-9

Crandall, R.W., Litan, R.E., Lehr, W. (2007). *The Effects of Broadband Deployment on Output and Employment: A Cross-Sectional Analysis of U.S. Data. Issues in Economic Policy.* The Brooklin Institution, Number 6.

Czernich, N. (2014). Does broadband internet reduce the unemployment rate? Evidence for Germany. *Information Economics and Policy*, *29*, 32–45. doi:10.1016/j.infoecopol.2014.10.001

Czernich, N., Falck, O., Kretschmer, T., & Woessmann, L. (2011). Broadband Infrastructure and Economic Growth. *Economic Journal (London)*, *121*(552), 505–532. doi:10.1111/j.1468-0297.2011.02420.x

Dutz, M. A., Orszag, J. M., & Willig, R. D. (2012). The liftoff of consumer benefits from the broadband revolution. *Review of Network Economics*, *11*(4). Advance online publication. doi:10.1515/1446-9022.1355

European Commission. (2010). A Digital Agenda for Europe. Brussels. Available at: https://www.kowi.de/Portaldata/2/Resources/fp/2010-com-digital-agenda.pdf

European Commission. (2016). Connectivity for a Competitive Digital Single Market – Towards a European Gigabit Society. COM(2016)587 final, Brussels.

European Commission. (2021). *2030 Digital Compass: The European way for the Digital Decade*. Communication from the Commission to the European Parliament, the Council, the European Economic and Social Committee and the Committee of Regions.

European Commission. (2022a). *Digital Economy and Society Index (DESI) 2022*. Thematic Chapters.

European Commission. (2022b). Digital Economy and Society Index (DESI) 2022. Austria. Brussels.

Eurostat. (2022). *Employed persons working from home as a percentage of the total employment*. Eurostat.

Fabling, R., & Grimes, A. (2021). Picking up speed: Does ultrafast broadband increase firm productivity? *Information Economics and Policy*, *57*, 100937. doi:10.1016/j.infoecopol.2021.100937

Federal Ministry for Transport, Innovation and Technology (2019). *Breitbandstrategie 2030*. Österreichs Weg in die Gigabit-Gesellschaft. Vienna. breitbandstrategie2030_ua (1).pdf

Ford, G. S. (2018). Is faster better? Quantifying the relationship between broadband speed and economic growth. *Telecommunications Policy*, *42*(9), 766–777. doi:10.1016/j.telpol.2018.05.006

Forman, C., Goldfarb, A., & Greenstein, S. (2012). The Internet and Local Wages: A Puzzle. *The American Economic Review*, *102*(1), 556–575. doi:10.1257/aer.102.1.556

Gillet, S. E., Lehr, W. H., Osario, C. A., & Sirbu, M. (2006). Measuring the Economic Impact of Broadband Deployment. *Final Report. National Technical Assistance, Training, Research and Evaluation Project No. 99-07-13829.*

Greenstein, S., & McDevitt, R. (2011). The Broadband Bonus: Estimating Broadband Internet's Economic Value. *Telecommunications Policy*, *35*(7), 617–632. doi:10.1016/j.telpol.2011.05.001

Holt, L., & Jamison, M. (2009). Broadband and contributions to economic growth: Lessons from the US experience. *Telecommunications Policy*, *33*(10-11), 575–581. doi:10.1016/j.telpol.2009.08.008

Ivus, O., & Boland, M. (2015). The employment and wage impact of broadband deployment in Canada. *The Canadian Journal of Economics. Revue Canadienne d'Economique*, *48*(5), 1803–1830. doi:10.1111/caje.12180

Jovanovic, B., & Rousseau, P. (2005). General Purpose Technologies. In Ph. Aghion & St. Durlauf (Eds.), *Handbook of Economic Growth* (Vol. 1, pp. 1181–1224). Part B.

Katz, R., & Callorda, F. (2020). *Assessing the Economic Potential of 10 G Networks*. Telecom Advisory Services.

Kolko, J. (2012). Broadband and Local Growth. *Journal of Urban Economics*, *71*(1), 100–113. doi:10.1016/j.jue.2011.07.004

Koutroumpis, P. (2009). The Economic Impact of Broadband on Growth: A Simultaneous Approach. *Telecommunications Policy*, *33*(9), 471–485. doi:10.1016/j.telpol.2009.07.004

Nordin, M., Grenestam, E., & Gullstrand, J. (2019). Is Super-fast Broadband Negative? An IV-Estimation of the Broadband Effect on Firms' Sales and Employment Level. School of Economics and Management. *Working Paper*, *2019*(8).

OECD. (2001). Understanding the Digital Divide. Digital Economy Papers. No. 49. Paris.

OECD. (2018). Bridging the Rural Digital Divide. OECD Digital Economy Papers. No. 265. Paris.

Qiang, C. Z.-W., & Rossotto, C. M. (2009). Economic impacts of broadband. In *Information and Communications for Development: Extending Reach and Increasing Impact*. World Bank, Global Information and Communication Technologies Department.

Ragnedda, M., & Muschert, G. W. (2013). *The Digital Divide. The Internet and Social Inequality in International Perspective*. Routledge. doi:10.4324/9780203069769

RTR (2022). *RTR Internet Monitor.* 2. Quartal 2022. lLOLnMDg5uSnEx67-report-q2-22-internet-monitor.pdf (23degrees.io)

Stanley, T. D., Doucouliagos, H., & Steel, P. (2018). Does ICT generate economic growth? A meta-regression analysis. *Journal of Economic Surveys, 32*(3), 705–726. doi:10.1111/joes.12211

Statistics Austria. (2021). Urban-Rural-Typologie von Statistik Austria. STATatlas.

Van de Werfhorst, H. G., Kessenich, E., & Geven, S. (2022). The digital divide in online education: Inequality in digital readiness of students and schools. *Computers and Education Online., 3*, 100100. doi:10.1016/j.caeo.2022.100100

Whitacre, B. E., Alam, M. R., & Lobo, B. J. (2018). Econometric error nullifies finding of the impact of broadband speed on county-level employment. *Information Economics and Policy, 44*, 58–60. doi:10.1016/j.infoecopol.2018.05.001

Whitacre, B. E., Gallardo, R., & Strover, S. (2014a). Does rural broadband impact jobs and income? Evidence from spatial and first-differenced regressions. *The Annals of Regional Science, 53*(3), 649–670. doi:10.100700168-014-0637-x

Whitacre, B. E., Gallardo, R., & Strover, S. (2014b). Broadband's contribution to economic growth in rural areas: Moving towards a causal relationship. *Telecommunications Policy, 38*(11), 1011–1023. doi:10.1016/j.telpol.2014.05.005

KEY TERMS AND DEFINITIONS

Broadband Adoption: Access to internet, the bandwidth defines the capacity how much data can be transmitted.

Broadband Policy: A policy adopted by governments to efficiently allocate resources for broadband roll-out.

Digital Divide: A gap between those who have access to (ultra-fast) internet and those who have no access (or only access to slow internet connections)

Digitalization Strategy: The European Union as well as nations and firms have formulated targets to manage the transition from analog to digital businesses as well as the transition from low broadband coverage to full coverage as well from low to ultra-fast speed broadband availability.

High-Speed Broadband Coverage: According to the goals set by the European Union, high-speed broadband coverage means the availability of internet speed of at least 30 Mbit/s.

Chapter 3
Empirical Evidence From Different Income Group Countries on the Interdependency Between Digitalization and Economic Development

Ajay Kumar Singh
https://orcid.org/0000-0003-0429-0925
DIT University, India

Bhim Jyoti
College of Forestry, Ranichauri, India

ABSTRACT

This study developed an information and communication technology evolution index (ICTEI) to explain the comparative performance of digitalization across 109 countries during 2010 – 2020. Composite Z-score and principal component analysis was used to create ICTEI. ICTEI was an integration of 16 different indicators associated with ICT access, ICT use, and ICT skills. Thereupon, ICTEI and per capita GDP were used as representative variables for digitalization and economic development, respectively to perceive a causal relationship among them. The empirical results revealed that economic development and digitalization have a positive and causal relationship. Trade openness, inflation, foreign direct investment, employment to population ratio, labour force participation rate, population growth, access to electricity, science and technological development, gross capital formation, and market exchange rate showed a diverse impact on economic development and digitalization in the different income groups countries.

DOI: 10.4018/978-1-6684-8634-4.ch003

INTRODUCTION

Digitalization is a process in which a country uses digital technologies, and information and communication technologies (ICTs), digital device and digital platform in production activities to be a green economy or digital economy (Jyoti & Singh, 2023). ICTs is a diverse combination of toots, technologies, methods and ways which help to disseminate and storge information, and exchange the information across communities. Digital technologies and ICTs are the significant drivers to create product and services, and helpful for a country to be digital economy (Benjamin & Woerter, 2019). Subsequently, any technology and device which enhance or create or develop an online platform for doing something known as digitalization. Digitalization helps to increase the productivity and efficiency of workers and firms (Dahlman et al., 2016), skills of workers, and create new livelihood opportunities for the people. Digital technologies also reduce transaction cost, and improve skills and knowledge of people (Nguyen, 2021). Digital technologies are positive to develop digital economy that produce positive influence on economic growth (Ma & Zhu, 2022; Li & Gospodarik, 2021; Jyoti & Singh, 2023). Furthermore, digital technologies have a vital contribution to increase scientific research outcome, create jobs and increase economic growth (Ogloblina et al., 2020; Dahmani et al., 2022; Singh et al., 2023d). Digital technologies are also potential to increase green economy and help to increase human well-being as sustaining the quality of ecosystem and natural resources. Thereupon, digitalization also creates a digital platform which may be supportive for producers and consumers to search desire information with minimum cost and time. Digitalization helps to increase the productivity, investment and growth of all sectors (Dahmani et al., 2022). Digitalization will be positive to reduce marginal cost and increase productivity of financial sector (Carbo-Valverde, 2017; Jyoti & Singh, 2023; Verma et al., 2023). ICT, artificial intelligence, digital device, GIS and weather forecasting software are useful for farmers to avoid the climate change risk and increase the production of agricultural sector (Singh et al., 2023e). In service sector, ICTs are also positive to increase quality of services and create opportunities for entrepreneurs and consumers (Aleksandrova et al., 2022). ICTs and digitalization also work as a significant driver to increase industrial development.

Further, internet network with high speed helps to increase the market accessibility for consumers and producers at minimum cost and enhance online trading of good and services (Kutsuri et al., 2019; Ogloblina et al., 2020; Soomro et al., 2022). Thus, internet facilities can be considered as a key determinant of economic growth and development (Afonasova et al., 2019; Irtyshcheva et al., 2021). Subsequently, internet is a crucial determinant of digitalization which promote e-commerce and e-business (Ogloblina et al., 2020). Henceforth, digitalization and ICTs indicators play a key role to increase appropriate business and entrepreneurship ecosystem (Carbo-Valverde,

2017). Digitalization is also supportive to create green entrepreneurship ecosystem which would increase green growth and green GDP (Singh et al., 2022b). In financial sector, digitalization and ICTs are helpful to improve the financial stability and increase transparency in money market. Consequently, digitalization creates employment, business opportunities, dissemination of information and communication across economic agents and countries (Dahlman et al., 2016). Past evidence indicates that digitalization would be positive to increase level of employment (Benjamin and Woerter, 2019; Ogloblina et al., 2020); skills of people through developing an online platform of learning; innovation in the market; perfect competition in market; digital market, online trading of goods and services; e-commerce and e-business (Ogloblina et al., 2020); price stability; minimize transportation and transaction charges; new business opportunities (Carbo-Valverde, 2017); foreign trade (exports and imports); discovery of new digital technologies; green technologies and appropriate technologies; environmentally friendly technologies; technology transfer and commercialization; financial accessibility, availability and stability (Carbo-Valverde, 2017); disseminate the information among the people; useability and accessibility of information for large users (i.e., producers, consumers and government); monitoring and transparency of government policies; and reduce high dependency of production sectors on natural resources. In environmental sector, digitalization and ICTs would be beneficial to abate GHGs and CO_2 emissions from production sector. Thus, it would be positive to reduce more possibilities of climate change and natural disaster in future. Singh et al. (2023d) have argued that digitalization helps to reduce environmental degradation. Consequently, it would promote green economy and green GDP, and increase natural resource sustainability (Singh et al., 2023d). Digitalization also has a positive impact on energy consumption (Bakry et al., 2023). It also increases the desired information of research and development (R&D) across scientific community across countries (Ogloblina et al., 2020). Thus, digitalization may be a crucial driver to increase science & technological development (S&TD) and innovation (Singh et al., 2019c). Accordingly, any country can be moved towards green economy as implementing the process of digitalization. Subsequently, use of digitalization, practices of green technologies, appropriate technologies, and digital technologies in production of goods and services would help to improve sustainable development and its associated pillars (i.e., economic, social, environmental and science & technological development) (Myovella et al., 2020; Aksentijevic et al., 2021; Ma & Zhu, 2022; Ashraf & Singh, 2022; Singh et al., 2022b).

In the stated perspectives, several studies theoretically and empirically proved that digitalization and ICT indicators are helpful to increase economic growth and development across countries (Raeskyesa & Lukas, 2019; Ogloblina et al., 2020; Li & Gospodarik, 2021). Existing studies have used different indicators like internet and mobile density, ICTs indicators, broadband internet, internet users and mobile subscription, individual using internet, fixed telephone subscriptions and mobile telecommunication to assess the impact of digitalization on economic growth and development (Habibi & Zabardast, 2020; Myovella et al., 2020; Nguyen, 2021; Sahoo et al., 2021; Soomro et al., 2022; Bakry et al., 2023; Singh et al., 2023c). Furthermore, previous studies have also created different indexes such as ICT development index, digital economy and society index, and information & communication technology index which were the integration of digitalization or digital economy enhancing factors (Afonasova et al., 2019; Özsoy et al., 2022). Subsequently, scientific research community used estimated indexed as dependent variables (*DV*s) and independent variables (*IV*s) in empirical models to assess the influence of digitalization on economic development (Farhadi et al., 2012; Li & Gospodarik, 2021; Jio & Sun, 2021). The empirical results of previous studies claimed that impact of digitalization and ICTs indicators on social-economic development is not uniform in different countries (Minges, 2015; Pradhan et al., 2022; Soomro et al., 2022). Few researchers could investigate the impact of digitalization on economic development in the group of different income countries (Jyoti & Singh, 2023). Despite that, there is a greater scope to analyse the impact of digitalization on economic development in different income group countries. Accordingly, the present chapter has formulated following hypothesis for further validation:

- Digitalization and economic development have a positive cause-and-effect relationship.
- The impact of digitalization and its determinants on economic development varies in different income group countries.
- The impact of economic development and its determinants on digitalization varies in different income group countries.

The present study has accomplished following objectives:

- To explain the progress of digitalization enhancing variables in high-income group countries (HIGCs), upper-middle-income group countries (UMIGCs), and lower-middle & low-income group countries (LMIGCs&LIGCs) during 2010 – 2020.
- To develop the information and communication technology evolution index (*ICTEI*) as an integration of 16 indicators using composite Z-score method and principal component analysis.
- To examine the impact of digitalization on economic development and vice-versa using a Cobb-Douglas production function.

The short map of this chapter is explained in Table 1.

Table 1. Short map of the chapter

Main Objectives
• To develop the information and communication technology evolution index (*ICTEI*) • To examine the impact of digitalization on economic development and vice-versa
Number of Countries and Time Period
109 countries from different income groups and 2010 – 2020
Source of Data
World Development Indicators (The World Bank)
Indicators of *ICTDI*
Fixed telephone subscriptions (per 100 people); mobile cellular subscriptions (per 100 people); secure internet servers (per 1 million people); ratio of ICT goods exports (% of total goods exports) with ICT goods imports (% total goods imports); ratio of communications, computer, etc. (% of service exports, BoP) with communications, computer, etc. (% of service imports, BoP); ratio of computer, communications and other services (% of commercial service exports) with computer, communications and other services (% of commercial service imports); fixed broadband subscriptions (per 100 people), individuals using the internet (% of population); automated teller machines (per 100,000 adults); education expenditure (% of GNI); compulsory education duration (years); and secondary school enrolment (% gross)
Method for *ICTDI* Estimation
Composite Z-score method and principal component analysis
Functional Form of the Variables
• $(ED) = f\{(ICTEI), (FDINI), (EGS), (ICP), (LFPRT), (EPR), (PGR), (AE), (STJA), (GCF)\}$ • $(ICTEI) = f\{(ED), (FDINI), (EGS), (IGDPD), (EPR), (PGR), (AE), (STJA), (MER)\}$
Empirical Models
• $ln(ED)_{ct} = \beta_0 + \beta_1\, ln(ICTEI)_{ct} + \beta_2\, ln(FDINI)_{ct} + \beta_3\, ln(EGS)_{ct} + \beta_4\, ln(ICP)_{ct} + \beta_5\, ln(LFPRT)_{ct} + \beta_6\, ln(EPR)_{ct} + \beta_7\, ln(PGR)_{ct} + \beta_8\, ln(AE)_{ct} + \beta_9\, ln(STJA)_{ct} + \beta_{10}\, ln(GCF)_{ct} + \theta_{ct}$ • $ln(ICTEI)_{ct} = \xi_0 + \xi_1\, ln(ED)_{ct} + \xi_2\, ln(FDINI)_{ct} + \xi_3\, ln(EGS)_{ct} + \xi_4\, ln(IGDPD)_{ct} + \xi_5\, ln(EPR)_{ct} + \xi_6\, ln(PGR)_{ct} + \xi_7\, ln(AE)_{ct} + \xi_8\, ln(STJA)_{ct} + \xi_9\, ln(MER)_{ct} + \mu_{ct}$
Principle Findings
• Digitalization, trade openness and S&TD have a positive impact on economic development. • Economic development, exports of goods and services, access to electricity and market exchange rate produced a positive impact on digitalization. • Economic development and digitalization have a positive and causal relationship.
Main Implications
• HIGCs should increase exports of goods and services, foreign direct investment, population growth and market exchange rate, employment to population ratio and S&TD to maintain digitalization and economic development. • UMIGCs and LMIGCs&LIGCs should control population growth and inflation to achieve a significant benefit from digitalization. • Better education and training for people will increase their capacity to use ICT indicators in production activities in developing countries. • LMIGCs&LIGCs should increase extensive investment in ICT and R&D ecosystem. • R&D expenditure, S&T development, technology transfer and commercialization, and strict IPRs regime help to increase digitalization and economic development in global countries.

REVIEW OF LITERATURE

Determinants of Digitalization and Economic Development

The concept of digitalization is not new in academic literature. Past studies have estimated the impact of digitalization on economic development by using diverse proxy variables across countries (Soomro et al., 2022). Carbo-Valverde (2017) claimed that digitalization is a revolution to nurture business ecosystem and maintain financial stability. Hernandez et al. (2016) argued that digital technology and economic growth has a positive correlation. Raeskyesa and Lukas (2019) reported a positive impact of ICTs indicators on economic growth. This study is also proposed that investment in human capital will be positive to increase ICT and transformation of global countries towards digital economy. Pradhan et al. (2022) have reported a diverse impact of ICTs indicators on economic growth in G-20 countries. Despite that, it argued that ICTs indicators and economic growth are mutually interdependent. Benjamin and Woerter (2019) have examined the consequence of digitalization in job creation in Switzerland. It found a positive impact of digitalization on job creation. Myovella et al. (2020) determined the implication of digitalization in economic growth in Sub Saharan Africa and OECD countries. Habibi and Zabardast (2020) have also found a positive impact of internet users and mobile subscription on economic growth in 10 - Middle East and 24- OECD countries. Singh et al. (2023d) have also noticed a positive impact of digitalization on environmental development.

Sahoo et al. (2021) have analysed the impact of digitalization and infrastructure on economic performance in India and China. Aksentijevic et al. (2021) have investigated the impact of ICTs on human development 130 countries. It noticed that ICTs has a positive impact on human development. Solomon & van Klyton (2020) also reported a positive impact of ICTs on economic growth in African countries. Nguyen (2021) have concluded that digitalization and governance have a positive impact on economic growth in 53 developing countries. Irtyshcheva et al. (2021) have estimated the impact of digital technologies on economic growth in Ukraine. Tiutiunyk et al. (2021) have noticed a bi-directional causality between digital transformation and stability in macroeconomic indicators across 10 European countries. Ma & Zhu (2022) have examined the impact of digital economy on high-quality green development in China. It observed a positive impact of digital economy on high-quality green development. Soomro et al. (2022) have also reported that ICTs help to increase economic growth in BRICS countries.

Interdependency of Economic Development With Digitalization

It is difficult to measure the progress of digitalization due to its complex association with social-economic development, science & technological development and exports and imports of ICTs related indicators. Therefore, a group of studies have created different indexes to investigate the progress of digitalization in global countries. Thereupon, previous studies have examined the influence of estimated indexes on economic development and growth. Farhadi et al. (2012) considered internet users, fixed broadband internet users and mobile users to create ICT index. Dobrota et al. (2012) have used 11 indicators associated with ICT access, ICT use and ICT skills to create ICT development index. Gerpott and Ahmadi (2015) have criticized the weightage procedure in ICT development index suggested by International Telecommunication Union (ITU). Ali et al. (2019) have developed modified ICT development index to avoid the discrepancy of previous studies in weightage estimation. In the study of Ali et al. (2019), modified ICT development index was the inclusion of affordability, efficiency and quality related variables. Afonasova et al. (2019) have developed digital economy and society index as a combination of ICT development index, global innovation index, network readiness index, percentage of internet users and high-technology exports in Russia and EU countries. Rath and Hermawan (2019) have also developed CT development index as an integration of telephone subscriptions, cellular phone subscription and fixed broadband subscription in Indonesia using PCA. Raeskyesa and Lukas (2019) have used ICT indicators (i.e., fixed-telephone subscriptions, mobile-cellular telephone subscriptions, international internet bandwidth per user, % of households with a computer, % of households with internet access) to measure the performance of digitalization in ASEAN countries.

Myovella et al. (2020) have also applied broadband internet and mobile telecommunication to notice the operation of digitalization in 74 Sub Saharan Africa (SSA) and OECD countries. Aly (2020) have considered digital adoption index, enabling digitalization index and digital evolution index to examine their association with economic development, labour productivity and employment rate in 25 developing countries. Tiutiunyk et al. (2021) have developed digital transformation index as an integration of digital readiness score, digital economy and society index, financial development index, global innovation index, digital competitiveness index, digital evolution index and ICT development index. Further, this study found a complex association between mentioned indicators and economic growth. Jio and Sun (2021) have used number of internet users per 100 people, computer service and software practitioners, total telecom business, number of mobile phone users per 100 people to create digital economy development index across Chinese cities. Zhang et al. (2021) have also generated digital development index as a combination of digital infrastructure, digital industry and digital integration for 30 cities in China. Appiah-Otoo and Song (2021) have created ICT index as an alignment of mobile users, internet and fixed broadband for 123 countries during 2002 – 2017.

Li and Gospodarik (2021) have developed digital economy development index in the Asia Pacific region using a factor analysis. This study has also observed the impact of digital economy development index on economic growth in China, Japan, South Korea, Brunei and Combodia. Aleksandrova et al. (2022) have employed index of digital life and digital adoption index to assess their interlinkages with per capita GDP in Russia. The findings of this study showed a positive impact of digital economic on economic growth. Özsoy et al. (2022) have created ICT development index as considering ICT access, usage and skills components of digitalization. Özsoy et al. (2022) have investigated the impact of ICT development index on technology intensity across countries. It inferred positive impact of digitalization on exports of high-tech products. Jyoti and Singh (2023) have investigated that impact of information and communication technology index (ICTI) on economic development and vice-versa. The findings of this study provide a confirmation the digitalization and economic development have a causal relationship. Verma et al. (2023) have developed ICT diffusion index to assess the impact of digitalization on financial development and economic development in 88 developing countries. Dahmani et al. (2022) have also observed the impact of ICT diffusion on economic growth and trade openness in Tunisia. Bakry et al. (2023) have examined the impact of ICT development index on growth and energy consumption in 27 countries.

RESEARCH METHOD AND MATERIALS

Inclusion of Countries

Total 109 countries from various income groups were finalized as per the availability of data related to ICT access, ICT use and ICT skills. The composite Z-score method was implemented to estimate country-wise *ICTEI*, while, ICTEI was the integration of 16 different indicators. The method for ICTEI estimation was adopted from the study of Jyoti and Singh (2023). This chapter comprised 43, 30, and 36 countries from high-income (HIGCs), upper-middle-income (UMIGCs) and lower-middle-income & low-income groups (LMIGCs&LIGCs), respectively. The strength of 109 countries in scientific and technical journal articles (*SRJA*), exports of goods and services (% of GDP) (*EGS*), per capita GDP (*ED*), foreign direct investment net inflows (% of GDP) (*FDINI*), inflation consumer prices (annual %) (*ICP*), inflation GDP deflator (annual %) (*IGDPD*), labour force participation rate total (*LFPRT*), employment to population ratio (*EPR*), gross capital formation (% of GDP) (*GCF*) and population growth (annual %) (*PGR*) is given in Table 2. The segregation of these countries in HIGCs, UMIGCs and LMIGCs&LIGCs is also given as:

HIGCs: Australia, Austria, Bahamas The, Belgium, Canada, Chile, Croatia, Cyprus, Czech Republic, Denmark, Estonia, Finland, France, Germany, Greece, Hungary, Iceland, Ireland, Israel, Italy, Japan, Korea Rep., Kuwait, Latvia, Lithuania, Luxembourg, Malta, Netherlands, New Zealand, Norway, Oman, Poland, Portugal, Saudi Arabia, Singapore, Slovak Republic, Slovenia, Spain, Sweden, Switzerland, United Kingdom, United States and Uruguay.

Table 2. Strength of 109 countries in main social-economic development related factors

Country	*SRJA*	*EGS*	*ED*	*FDINI*	*ICP*	*IGDPD*	*LFPRT*	*EPR*	*GCF*	*PGR*
Albania	180	23.1	4390	7.19	1.6	0.45	57.8	50.1	22.15	0.57
Algeria	5231	18.0	3834	0.78	2.4	5.51	39.9	34.9	43.38	1.84
Armenia	521	29.8	4021	0.37	1.2	2.02	51.3	40.4	18.53	0.19
Australia	53610	24.0	58030	1.48	0.8	1.78	65.0	60.8	22.27	1.28
Austria	12362	51.4	43346	4.20	1.4	2.31	60.8	57.5	25.90	0.42
Azerbaijan	761	35.7	5083	1.19	2.8	7.58	62.8	58.7	24.28	0.68
Bahamas, The	20	24.3	27209	3.63	0.0	11.96	66.4	57.6	22.65	0.96
Bangladesh	3135	12.0	1644	0.35	5.7	4.09	56.7	53.6	30.47	1.00
Belarus	1180	61.9	6235	2.31	5.5	10.10	64.0	61.0	26.27	0.42
Belgium	15688	80.0	40439	3.37	0.7	1.28	53.8	50.8	24.16	0.48
Benin	228	19.9	1215	1.12	3.0	2.88	70.8	69.7	25.63	2.69
Bolivia	103	20.4	2983	3.05	0.9	1.91	64.5	59.4	15.84	1.38
Brazil	60148	16.9	8229	2.62	3.2	4.81	57.3	49.4	15.40	0.71
Bulgaria	3311	55.3	7921	3.87	1.7	4.20	55.5	52.7	18.75	0.60
Burkina Faso	252	31.8	732	0.83	1.9	6.84	64.8	61.6	24.15	2.82
Cameroon	876	15.0	1420	1.65	2.4	0.55	75.3	72.4	17.74	2.55
Canada	59968	29.4	42259	1.61	0.7	0.75	64.0	58.0	22.26	1.15
Chile	7122	31.5	12954	3.37	3.0	8.36	55.3	49.1	19.82	0.86
China	528263	18.5	10370	1.44	2.4	0.63	68.3	64.8	43.54	0.23
Colombia	7195	13.5	5892	2.75	2.5	1.41	62.7	53.3	19.01	1.07
Costa Rica	507	31.5	12106	3.40	0.7	0.20	58.7	48.5	16.56	0.92
Cote d'Ivoire	248	21.6	2314	0.83	2.4	0.97	55.4	53.5	22.13	2.54
Croatia	4277	42.0	12985	2.06	0.2	0.13	51.0	47.2	23.91	0.43
Cyprus	1245	76.0	26373	1.33	0.6	1.18	62.7	57.9	18.87	0.73
Czech Republic	15577	71.0	18985	2.62	3.2	4.40	59.8	58.3	25.92	0.24
Denmark	13979	54.9	56202	0.44	0.4	2.61	62.0	58.5	22.93	0.29
Dominican Republic	49	18.3	7678	3.11	3.8	4.72	60.3	56.7	25.38	1.01
Ecuador	2142	22.5	5318	1.16	0.3	0.92	60.1	56.4	23.12	1.54
Egypt Arab Republic	13327	13.1	4028	1.60	5.0	6.22	41.2	37.4	13.79	1.92
El Salvador	45	25.5	3632	1.21	0.4	0.20	55.5	52.0	17.91	0.50
Estonia	1415	71.2	19767	11.54	0.4	0.29	63.6	59.3	30.24	0.20
Finland	10599	36.2	44779	0.88	0.3	1.24	59.0	54.4	24.46	0.14
France	66352	27.9	35786	0.50	0.5	2.52	54.5	50.2	23.81	0.19
Gambia, the	34	8.0	692	10.15	5.9	6.26	56.8	50.5	33.18	2.90
Georgia	550	37.3	4448	3.37	5.2	7.28	59.3	48.3	23.86	0.07

continued on following page

Table 2. Continued

Country	*SRJA*	*EGS*	*ED*	*FDINI*	*ICP*	*IGDPD*	*LFPRT*	*EPR*	*GCF*	*PGR*
Germany	104396	43.4	41315	3.71	0.5	1.60	61.6	59.3	21.15	0.08
Ghana	1276	32.2	2019	2.74	9.9	7.11	68.4	65.2	19.49	2.13
Greece	10907	32.0	17324	1.75	1.2	0.84	51.0	42.7	15.00	0.20
Guatemala	100	16.9	4126	1.10	3.2	2.65	56.8	54.8	12.93	1.52
Honduras	45	35.2	2223	1.00	3.5	4.63	59.5	54.5	18.72	1.61
Hungary	6701	79.5	14369	108.42	3.3	5.92	56.9	54.5	27.49	0.22
Iceland	681	34.2	53188	4.73	2.8	3.21	72.3	68.3	21.58	1.62
India	135788	18.7	1812	2.42	6.6	4.61	44.9	41.3	29.28	0.99
Indonesia	26948	17.2	3757	1.81	1.9	0.46	67.4	64.5	32.37	1.07
Ireland	7174	131.1	78733	7.62	0.3	1.21	61.1	57.7	40.88	1.03
Israel	12235	28.0	37488	5.96	0.6	0.97	61.8	59.1	22.12	1.76
Italy	71240	29.4	29360	1.17	0.1	1.17	48.5	44.1	17.58	0.47
Jamaica	164	37.9	4539	1.92	5.2	3.56	61.5	55.7	24.41	0.44
Japan	98793	15.5	34813	1.24	0.0	0.84	62.0	60.3	25.51	0.34
Jordan	2627	23.7	4029	1.64	0.3	0.26	37.8	30.6	11.94	1.00
Kazakhstan	2367	30.2	10975	4.33	6.7	4.21	69.2	65.8	28.52	1.29
Kenya	1247	9.7	1560	0.71	5.4	5.18	73.1	68.9	19.67	2.25
Korea, Rep.	66376	36.4	31327	0.56	0.5	1.31	62.8	60.3	31.86	0.14
Kuwait	1004	61.6	24429	0.59	2.1	14.07	68.9	66.5	25.06	1.50
Kyrgyz Republic	137	31.3	1100	5.19	6.3	5.76	55.9	51.0	29.01	1.90
Latvia	1418	60.3	15584	2.80	0.2	0.09	61.7	56.7	22.36	0.70
Lesotho	19	47.1	983	1.58	5.0	0.03	63.2	47.7	27.74	0.80
Lithuania	2267	73.5	17214	7.92	1.2	1.46	62.6	57.3	13.47	0.03
Luxembourg	869	204.7	104879	84.72	0.8	4.27	60.8	56.7	17.90	1.67
Madagascar	127	20.1	442	2.75	4.2	4.35	84.1	82.0	18.24	2.64
Malaysia	23661	61.4	10632	1.28	1.1	0.78	64.9	62.0	19.73	1.29
Mali	90	29.2	782	1.76	0.4	0.53	68.4	63.1	17.17	2.97
Malta	422	140.6	25006	29.95	0.6	1.45	62.3	59.7	23.42	2.21
Mauritius	127	30.0	9058	2.25	2.6	1.33	56.5	52.3	18.20	0.00
Mexico	16346	40.2	8910	2.89	3.4	2.91	55.5	53.0	19.31	1.06
Moldova	210	26.5	3250	1.32	3.8	5.45	40.3	38.8	23.43	1.68
Mongolia	141	57.7	4127	12.91	3.7	3.71	58.8	54.7	22.38	1.63
Morocco	5057	34.9	2819	1.54	0.7	0.86	43.2	38.2	28.45	1.20
Myanmar	231	28.3	1587	2.30	1.4	5.06	56.7	56.1	29.66	0.67
Namibia	156	33.4	4048	1.53	2.2	5.82	58.0	45.6	14.40	1.84
Nepal	792	6.8	1028	0.38	5.1	3.61	77.4	73.7	28.40	1.83

continued on following page

Table 2. Continued

Country	*SRJA*	*EGS*	*ED*	*FDINI*	*ICP*	*IGDPD*	*LFPRT*	*EPR*	*GCF*	*PGR*
Netherlands	30457	77.9	46345	16.34	1.3	2.29	64.6	62.1	21.74	0.56
New Zealand	7889	21.8	40218	1.93	1.7	1.29	70.3	67.0	22.41	2.09
Nicaragua	44	42.3	1922	1.44	3.7	5.86	63.1	59.3	19.18	1.20
Nigeria	5602	8.8	2396	0.55	11.5	7.85	53.4	48.3	29.40	2.54
Norway	11803	32.2	75017	1.05	1.3	3.60	63.8	61.0	30.31	0.59
Oman	856	44.2	15743	3.87	0.9	13.22	67.0	65.0	20.75	2.61
Pakistan	12904	10.0	1447	0.80	9.7	10.14	49.9	47.7	15.33	1.98
Panama	173	39.6	12172	3.99	1.6	1.80	57.9	50.5	24.06	1.60
Paraguay	98	33.5	5671	1.75	1.8	2.68	70.1	64.8	20.16	1.24
Peru	1630	22.3	5792	1.48	2.0	4.21	65.3	60.6	17.82	1.41
Philippines	2237	25.2	3270	1.82	2.6	1.64	54.8	53.4	17.38	1.35
Poland	35663	56.2	14661	2.91	3.4	4.10	56.1	54.3	17.25	0.18
Portugal	14295	37.0	19772	2.06	0.0	1.94	58.0	54.1	18.75	0.11
Romania	10345	37.3	10845	1.45	2.6	3.75	55.1	52.3	24.51	0.59
Russian Federation	81579	25.5	9667	0.64	3.4	0.90	61.9	58.4	23.99	0.21
Rwanda	170	19.0	834	0.97	9.9	8.27	82.2	80.9	24.53	2.54
Saudi Arabia	10898	26.3	18691	0.77	3.4	7.93	61.0	56.5	26.31	1.58
Senegal	388	20.6	1365	6.01	2.5	2.34	44.3	42.7	30.51	2.71
Singapore	11459	176.2	58057	25.72	0.2	2.92	69.1	66.3	22.62	0.31
Slovak Republic	5322	85.4	17361	0.31	1.9	2.37	59.0	55.1	19.35	0.09
Slovenia	3206	77.9	22899	0.91	0.1	1.23	57.8	54.9	20.00	0.67
South Africa	13009	27.8	5659	0.95	3.2	5.27	50.5	35.8	12.75	1.27
Spain	54537	30.6	24939	2.63	0.3	1.10	56.7	47.9	20.69	0.48
Sri Lanka	1348	16.6	4053	0.54	6.2	3.43	48.5	45.6	25.24	0.53
Sudan	398	0.4	1128	3.36	150.3	115.28	47.6	38.3	31.93	2.39
Sweden	20421	44.6	51540	5.25	0.5	1.72	64.4	59.0	24.79	0.72
Switzerland	21379	62.3	85685	34.21	0.7	0.50	67.9	64.6	28.36	0.71
Tanzania	603	14.3	1061	1.10	3.3	0.34	83.1	80.9	41.02	2.94
Thailand	12514	51.5	6198	0.97	0.8	1.07	67.0	66.3	23.88	0.25
Togo	86	21.8	627	8.44	1.8	1.28	57.3	55.1	23.61	2.40
Tonga	5	19.0	4630	0.86	0.3	3.62	46.3	44.6	24.82	1.14
Tunisia	5565	43.6	3781	1.42	5.6	5.07	46.0	38.4	22.38	1.05
Turkey	33536	28.7	12039	1.06	12.3	14.83	49.3	42.8	31.89	1.08
Ukraine	10380	39.1	2344	0.20	2.7	9.79	55.1	50.1	7.52	0.57
United Kingdom	97681	28.1	43020	1.13	1.0	5.35	62.8	60.0	16.70	0.57

continued on following page

Table 2. Continued

Country	*SRJA*	*EGS*	*ED*	*FDINI*	*ICP*	*IGDPD*	*LFPRT*	*EPR*	*GCF*	*PGR*
United States	422808	10.1	58203	1.01	1.2	1.21	60.9	56.0	21.09	0.96
Uruguay	852	25.4	15045	0.56	9.8	10.87	61.0	54.6	17.00	0.35
Zimbabwe	359	37.2	1307	0.83	557.2	558.56	83.4	79.0	7.45	1.47

Source: World Development Indicator (World Bank) (2020).

UMIGCs: Albania, Armenia, Azerbaijan, Belarus, Brazil, Bulgaria, China, Colombia, Costa Rica, Dominican Republic, Ecuador, Georgia, Guatemala, Jamaica, Jordan, Kazakhstan, Malaysia, Mauritius, Mexico, Moldova, Namibia, Panama, Paraguay, Peru, Romania, Russian Federation, South Africa, Thailand, Tonga and Turkey.

LMIGCs&LIGCs: Algeria, Bangladesh, Benin, Bolivia, Cameroon, Cote d'Ivoire, Egypt Arab Republic, El Salvador, Ghana, Honduras, India, Indonesia, Kenya, Kyrgyz Republic, Lesotho, Mongolia, Morocco, Myanmar, Nepal, Nicaragua, Nigeria, Pakistan, Philippines, Senegal, Sri Lanka, Tanzania, Tunisia, Ukraine, Zimbabwe, Burkina Faso, Gambia The, Madagascar, Mali, Rwanda, Sudan and Togo.

Data Sources and Time Period of Country-Wise Panel

The statistics of related variables were derived from the official website of world development indicators (The World Bank). Furthermore, data of most indicators were available for the time period, 2010 – 2020. Therefore, this chapter could compile the panel of identified indicators for each income groups of countries.

Consistency of Dependent and Independent Variables

This chapter examines the causal association among the digitalization and economic development in different income group countries. For stated investigation, per capita gross domestic product and information communication technology evolution index (*ICTEI*) were used as proxy variables for economic development and digitalization, respectively in this chapter. Prior studies claimed that per capita gross domestic product (*PCGDP*) reflects the overall well-being, social development and livelihood security of people. Earlier studies were also preferred per capita GDP to observe the interconnection of economic development with entrepreneurship ecosystem, foreign direct investment, employment rate, urbanization, population growth, inflation, consumer price inflation, market exchange rate, digitalization, ICT import and exports, capital formation, foreign trade and trade openness (Singh et al., 2019a;

Singh & Ashraf, 2020; Nguyen, 2021; Singh & Kumar, 2022a; Singh et al., 2023a). Moreover, past studies used different indicators to determine the implications of digitalization on economic development and growth, human development, job creation, labour productivity, financial stability and exports of high-tech goods (Carbo-Valverde, 2017; Benjamin & Woerter, 2019; Myovella et al., 2020; Zhang et al., 2021; Aksentijevic et al., 2021; Özsoy et al., 2022; Aleksandrova et al., 2022). Also, few studies developed various indexes to measure the progress of digitalization across countries (Jyoti & Singh, 2023). Subsequently, most studies considered estimated indexes as independent variables to examine the integrated influence of digitalization on economic development. Existing researchers have used various indexes to examine the causal interlinkages of digitalization with economic, financial, social and environmental development (Singh et al., 2023d; Verma et al., 2023). In this chapter, we also create *ICTEI* as a composition of 16 variables that are positive to increase digitalization for 109 countries using a principal component analysis (Hermawan, 2019; Singh & Kumar, 2022b; Jyoti & Singh, 2023). *ICTEI* was the inclusion of fixed telephone subscriptions (per 100 people); mobile cellular subscriptions (per 100 people); secure internet servers (per 1 million people); ratio of ICT goods exports (% of total goods exports) with ICT goods imports (% total goods imports); ratio of communications, computer, etc. (% of service exports, BoP) with communications, computer, etc. (% of service imports, BoP); ratio of computer, communications and other services (% of commercial service exports) with computer, communications and other services (% of commercial service imports); fixed broadband subscriptions (per 100 people); individuals using the internet (% of population); automated teller machines (per 100,000 adults); education expenditure (% of GNI); compulsory education duration (years); and secondary school enrolment (% gross). Accordingly, *PCGDP* and *ICTEI* were considered as dependent and independent variables in regression analysis in this chapter.

Furthermore, previous studies have reported that foreign direct investment net inflows, exports of goods and services, inflation consumer prices, inflation GDP deflator, labour force participation rate, employment to population ratio, population growth, access to electricity, scientific and technical journal articles, gross capital formation and market exchange rate have a significant impact on economic development and digitalization (Sahoo et al., 2021; Jyoti & Singh, 2023). Thus, these variables were also used as independent variables in the regression analysis. Moreover, science & technological development (S&TD) has a positive influence on digital technologies and devices which increase digitalization. S&TD is also beneficial to meet the technological requirements of manufacturing firms which produce innovative goods and services, and create new business and jobs opportunities (Singh et al., 2019c; Singh et al., 2020a; Singh & Jyoti, 2020; Singh & Kumar, 2021). Scientific and technical journal articles help to disseminate technology, knowledge and innovation among the scientific community and manufacturing firms. So, scientific and technical journal articles was used to capture the impact of S&TD on digitalization and economic development (Singh & Kumar, 2022b).

Empirical Analysis

Earlier studies used different models to examine the interconnection between estimated indexes in different areas. Kumar and Sharma (2013); Kumar et al. (2017); Singh et al. (2017); Sharma and Singh (2017); Singh et al. (2022c) have observed the implications of climate change on food security index in India using a linear, non-linear and log-linear regression models. Kumar et al. (2015) have assessed the impact of carbon emissions on global food security index across countries using a linear regression model. Singh and Issac (2018) have considered linear, non-linear and log-linear regression equations to assess the implications of climate change on sustainable livelihood security index in Gujarat, India. Singh et al. (2019a) have applied a linear regression model to detect the association between environmental sustainability index, human development index, social and economic development index across countries. Ashraf and Singh (2019); Singh and Ashraf (2020) have adopted log-linear regression equations to explore the influence of entrepreneurship ecosystem index on economic development across countries. Singh et al. (2020a) have adopted different forms of empirical models to examine the interconnection among the intellectual property rights index, science & technological development index and manufacturing sector in selected countries. Singh et al. (2020b) have observed the inter-causal association between economic, social, environmental and science & technological development, and global sustainable development. Tiutiunyk et al. (2021) have investigated the causal association between digital transformation index, financial development index, global innovation index, digital competitiveness index, digital evolution index and ICT development index in Ukraine. Singh et al. (2021); Singh and Kumar (2022b) have also included GSDI, EDI and SDI to explain causal association among them using robust empirical models. Aksentijevic et al. (2021) have assessed the impact of ICT on human development index across 130 countries. Singh et al. (2022a) have applied numerous systems of regression equations to explore the impact of climatic factors on agricultural sustainability index in Indian states. Aleksandrova et al. (2022) have investigated the relationship between global competitiveness index, index of digital life, digital adoption index and resilience index in Russia using a descriptive analysis. Singh and Kumar (2022a); Singh et al. (2023a) have assessed the causal association between entrepreneurship ecosystem index and economic development using log-linear regression analysis across countries. Singh et al. (2022b) have considered linear regression equations to examine the interconnection between GSDI, EDI, SDI, EEI and GEEI across countries. Singh et al. (2023b) have applied log-linear functional form of regression equations to observe the causal association of GEEI with sustainable development across countries.

Above-mentioned literature stated that estimated indexes can be included as dependent and independent variables in the empirical models. In this chapter, therefore, *ICTEI* was applied as a dependent and independent variable to observe its causal association with economic development in HIGCs, UMIGCs, and LMIGCs&LIGCs. It used a log-linear functional form of regression model (Solomon & van Klyton, 2020). Following regression equations was employed to examine the impact of *ICTEI* and explanatory variables on economic development:

$$ln(ED)_{ct} = \beta_0 + \beta_1 ln(ICTEI)_{ct} + \beta_2 ln(FDINI)_{ct} + \beta_3 ln(EGS)_{ct} + \beta_4 ln(ICP)_{ct} + \beta_5 ln(LFPRT)_{ct} + \beta_6 ln(EPR)_{ct} + \beta_7 ln(PGR)_{ct} + \beta_8 ln(AE)_{ct} + \beta_9 ln(STJA)_{ct} + \beta_{10} ln(GCF)_{ct} + \theta_{ct} \quad (1)$$

Here, *ED* is economic development; *ICTEI* is information and communication technology evolution index; *c* is cross-section country; *t* is time period; *ln* is natural logarithm of respective variables; β_0 is constant coefficient that measure the total factor productivity (*TFP*); $\beta_1, \beta_2, \ldots, \beta_{10}$ are the elasticities of respective variables; θ_{ct} is residual term in equation (1). The explanation of remaining variables is given in Table 3. The elasticities measure the percentage change in economic development as 1% change in the values of associated variables. Following regression model was included to assess the influence of economic development and other independent variables on *ICTEI*:

$$ln(ICTEI)_{ct} = \xi_0 + \xi_1 ln(ED)_{ct} + \xi_2 ln(FDINI)_{ct} + \xi_3 ln(EGS)_{ct} + \xi_4 ln(IGDPD)_{ct} + \xi_5 ln(EPR)_{ct} + \xi_6 ln(PGR)_{ct} + \xi_7 ln(AE)_{ct} + \xi_8 ln(STJA)_{ct} + \xi_9 ln(MER)_{ct} + \mu_{ct} \quad (2)$$

Here, *ICTEI* is information and communication technology evolution index; *ED* is economic development; ξ_0 is constant coefficient that measure the *TFP*; $\xi_1, \xi_2, \ldots, \xi_9$ are the elasticities of respective variables; μ_{ct} is error-term in equation (2). $\xi_1, \xi_2, \ldots, \xi_9$ measure the percentage variation in digitalization due to 1% increase in the values of *IVs*. The description of remaining variables is given in Table 3. In this chapter, we used three panels i.e., HIGCs, UMIGCs and LMIGCs&LIGCs to examine the impact of *IVs* on economic development and digitalization in each income group countries. SPSS and STATA statistical software were used to estimate the regression coefficient of *IVs* with *ED* and *ICTEI*. While, coefficients of *IVs* with *ED* and *ICTEI* were assessed through feasible generalized least square estimation as assuming that panel data of HIGCs, UMIGCs, and LMIGs&LIGCs have autocorrelation, serial correlation and heteroskedasticity. Past studies also

Table 3. Brief description of variables

Indicators	Unit	Symbol
Information and communication technology evolution index	Number	*ICTEI*
Exports of goods and services (% of GDP)	%	*EGS*
GDP per capita (constant 2015 US$)	US$	*ED*
Foreign direct investment net inflows (% of GDP)	%	*FDINI*
Inflation consumer prices (annual %)	%	*ICP*
Inflation GDP deflator (annual %)	%	*IGDPD*
Labour force participation rate total (% of total population ages 15+)	%	*LFPRT*
Employment to population ratio (15+, total (%)	%	*EPR*
Gross capital formation (% of GDP)	%	*GCF*
Population growth (annual %)	%	*PGR*
Access to electricity (% of population)	%	*AE*
Price level ratio of PPP conversion factor (GDP) to market exchange rate	%	*MER*
Scientific and technical journal articles	Number	*STJA*

Source: Author's compilation based on existing studies.

applied above-mentioned mentioned model to estimate the regression coefficient of *IV*s in the existence of autocorrelation, serial correlation and heteroskedasticity in the country-wise panel data analysis (Kumar et al., 2015; Kumar et al., 2017; Pradhan et al., 2022; Jyoti and Singh, 2023).

RESULTS AND DISCUSSION

Performance of ICT Enhancing Factors in the HIGCs, UMIGCs, and LMIGCs&LIGCs

It is stated earlier that ICT access, ICT use and ICT skills related variables help to create a conducive ecosystem of digitalization (Jyoti and Singh, 2023). Therefore, trend in variables associated with digitalization for the time period, 2010 – 2020 is presented here:

Trend in ICT Access Enhancing Factors: Trend in fixed telephone subscriptions (per 100 people), mobile cellular subscriptions (per 100 people), secure internet servers (per 1 million people), ICT goods exports (% of total goods exports), ICT goods imports (% total goods imports), ICT service exports (% of service exports, BoP), communications, computer, etc. (% of service exports, BoP) and communications, computer, etc. (% of service imports, BoP) for HIGC, LIGC, LMIGC and UMIGC are given in Figure 1, Figure 2, Figure 3, Figure 4, Figure 5, Figure 6, Figure 7 and Figure 8, respectively. These indicators help to increase ICT accessibility. The group of HIGCs and UMIGCs have better performance, and LIGCs and LMIGCs have relatively lower position in aforesaid indicators.

Trend in ICT Use Enhancing Factors: The trend in fixed broadband subscriptions (per 100 people); individuals using the internet (% of population); Automated teller machines (ATMs) (per 100,000 adults); Computer, communications and other services (% of commercial service exports); and computer, communications and other services (% of commercial service imports) for HIGCs, LIGCs,

Figure 1. Fixed telephone subscriptions (per 100 people)
Source: World Development Indicators (WDI) (World Bank).

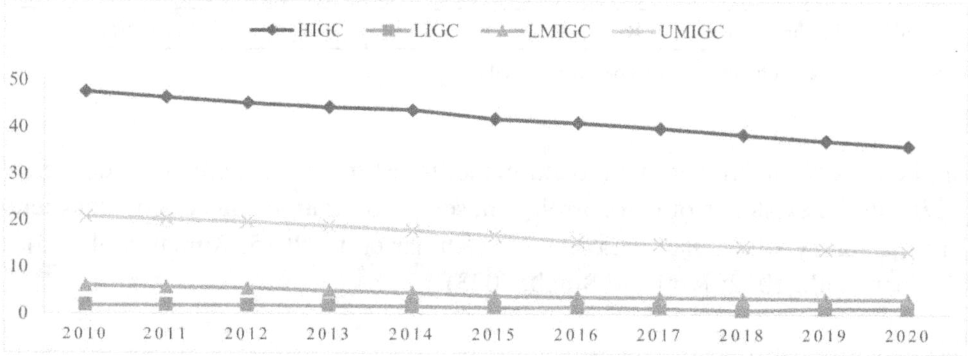

Figure 2. Mobile cellular subscriptions (per 100 people)
Source: WDI (World Bank).

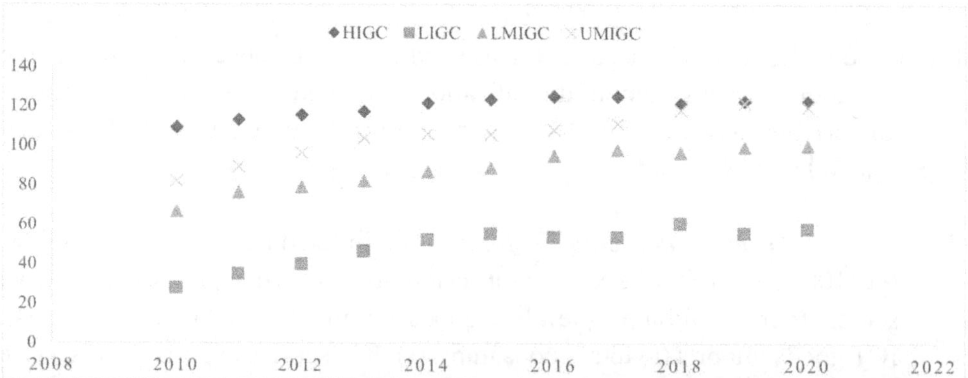

Figure 3. Secure Internet servers (per 1 million people)
Source: WDI (World Bank).

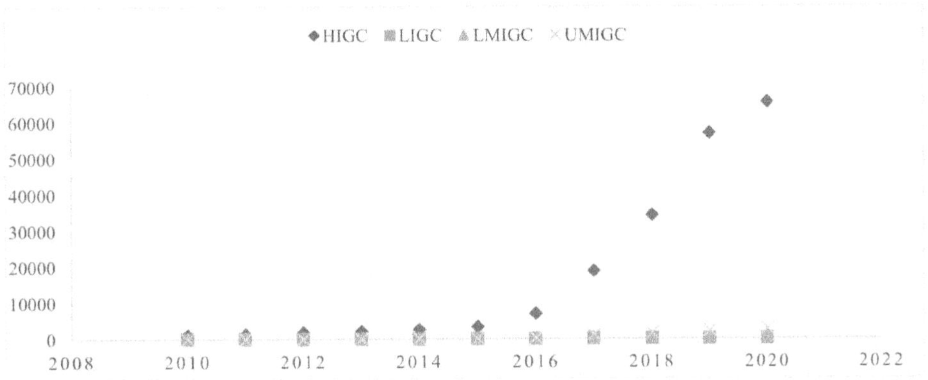

Figure 4. ICT goods exports (% of total goods exports)
Source: WDI (World Bank).

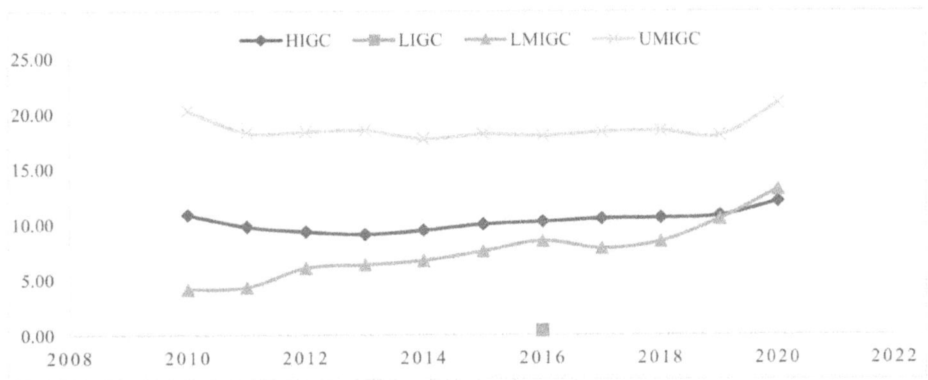

Figure 5. ICT goods imports (% total goods imports)
Source: WDI (World Bank).

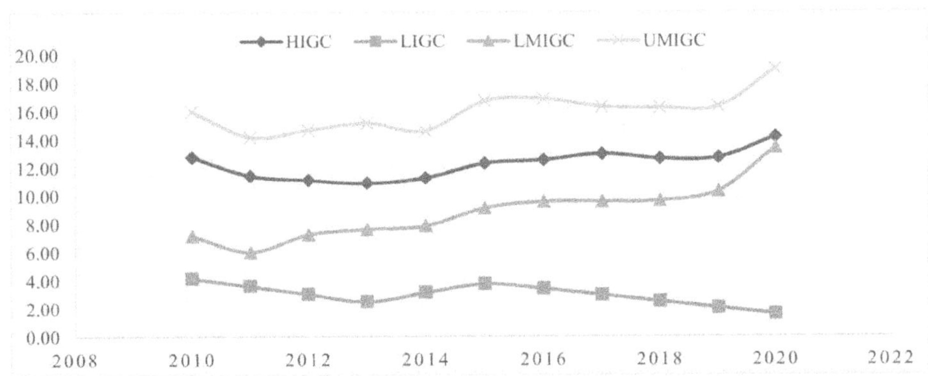

Figure 6. ICT service exports (% of service exports, BoP)
Source: WDI (World Bank).

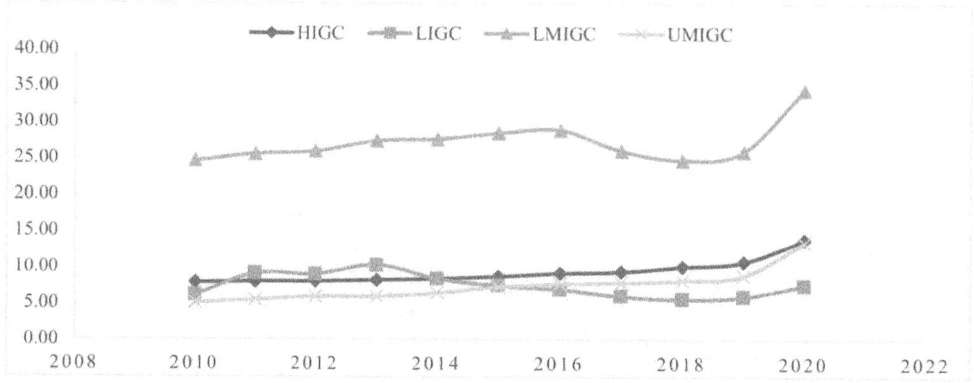

Figure 7. Communications, computer, etc. (% of service exports, BoP)
Source: WDI (World Bank).

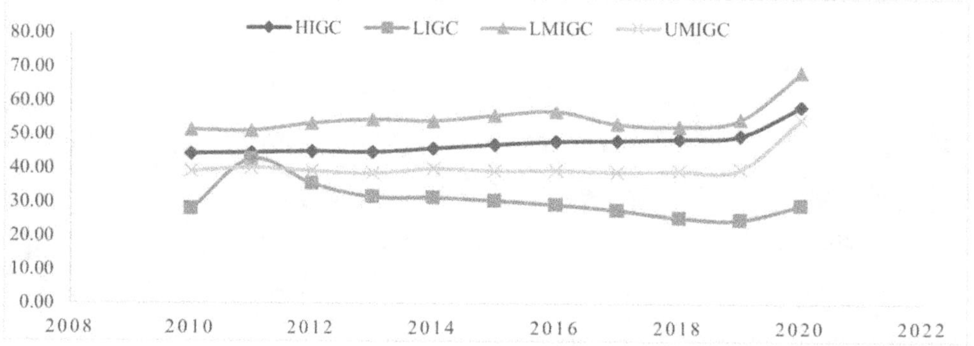

Figure 8. Communications, computer, etc. (% of service imports, BoP)
Source: WDI (World Bank).

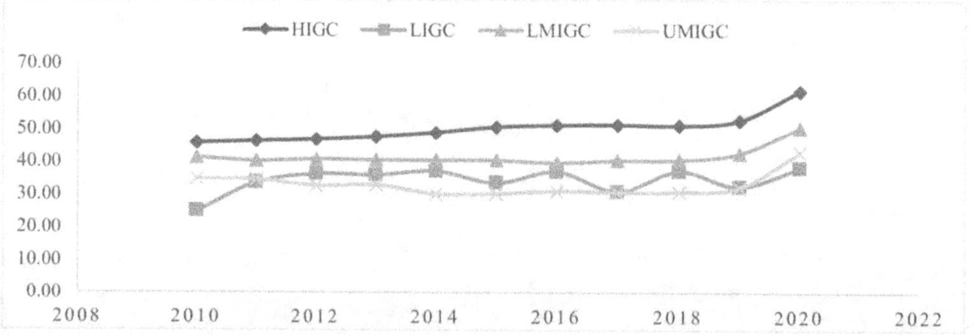

LMIGC and UMIGC are given in Figure 9; Figure 10; Figure 11; Figure 12; and Figure 13, respectively.

Trend in ICT Skills Enhancing Factors: Trend in education expenditure (% of GNI), compulsory education duration (years) and secondary school enrollment (% gross) for HIGCs, LIGCs, LMIGCs and UMIGCs are given in Figure 14, Figure 15 and Figure 16, respectively. The progress of most indicators of ICT access, ICT use and ICT skills in HIGCs are continued to be increased during 2010-2022. Thus, high-income group countries have a better position to create an appropriate infrastructure of digitalization. Also, UMIGCs have an increasing trend in most indicators which are positive to create a favorable ecosystem of digitalization. On the other hand, most indicators of ICT skills have a positive and increasing trend in LMIGCs and LIGCs. Thus, it infers that

Figure 9. Fixed broadband subscriptions (per 100 people)
Source: WDI (World Bank).

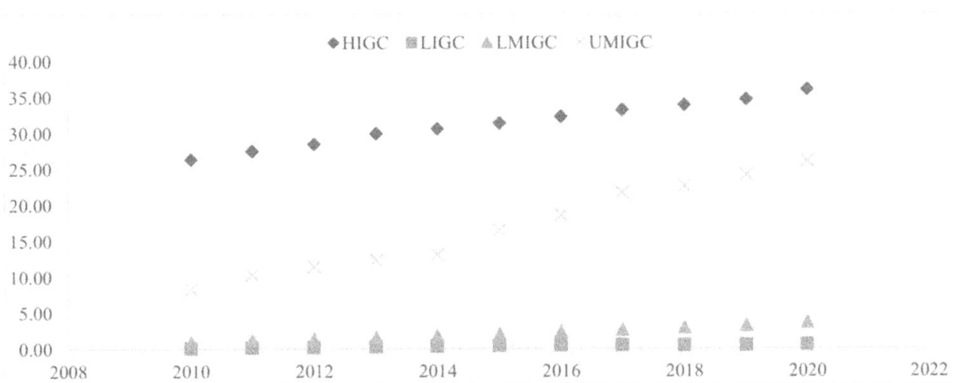

Figure 10. Individuals using the internet (% of population)
Source: WDI (World Bank).

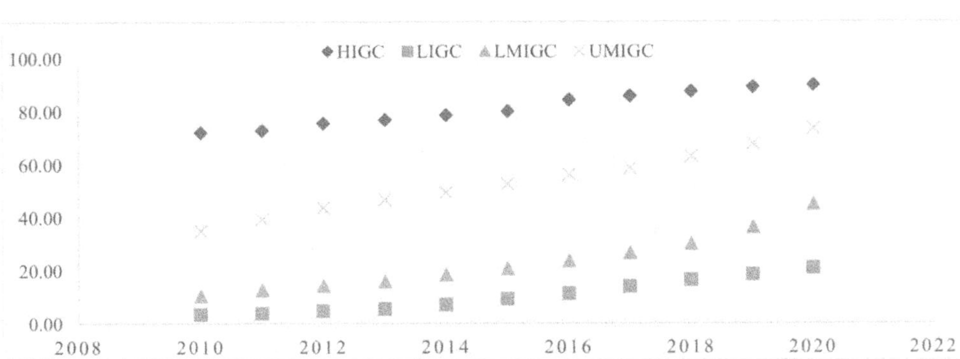

Figure 11. Automated teller machines (ATMs) (per 100,000 adults)
Source: WDI (World Bank).

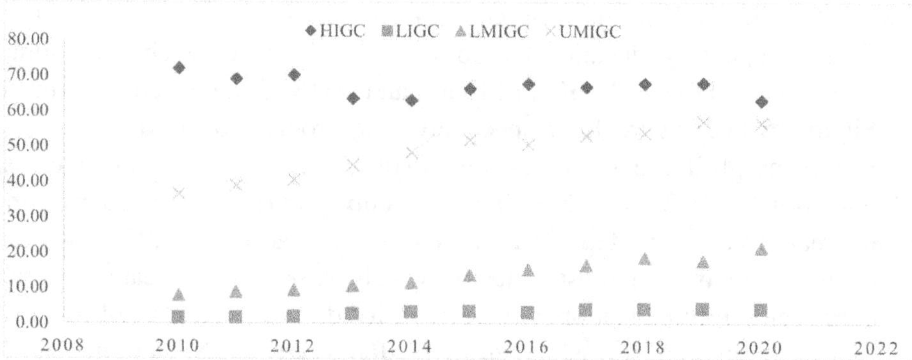

Figure 12. Computer, communications and other services (% of commercial service exports)
Source: WDI (World Bank).

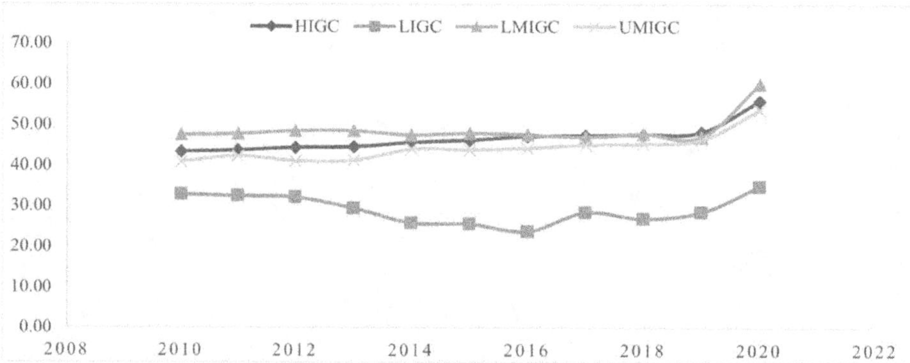

Figure 13. Computer, communications and other services (% of commercial service imports)
Source: WDI (World Bank).

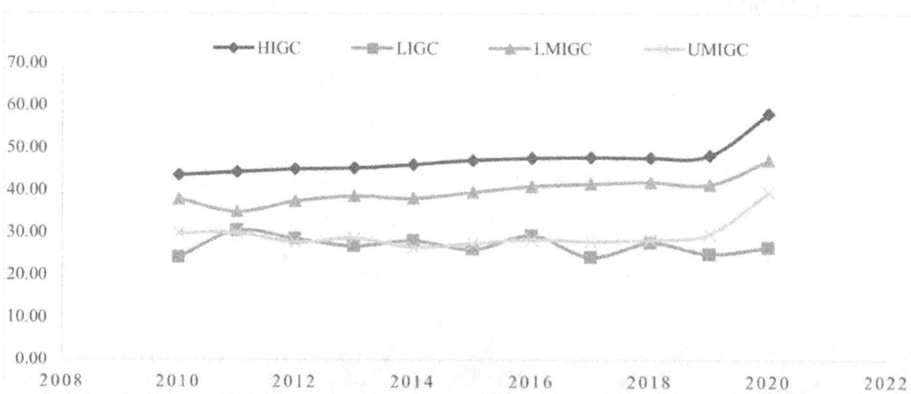

Figure 14. Education expenditure (% of GNI)
Source: WDI (World Bank).

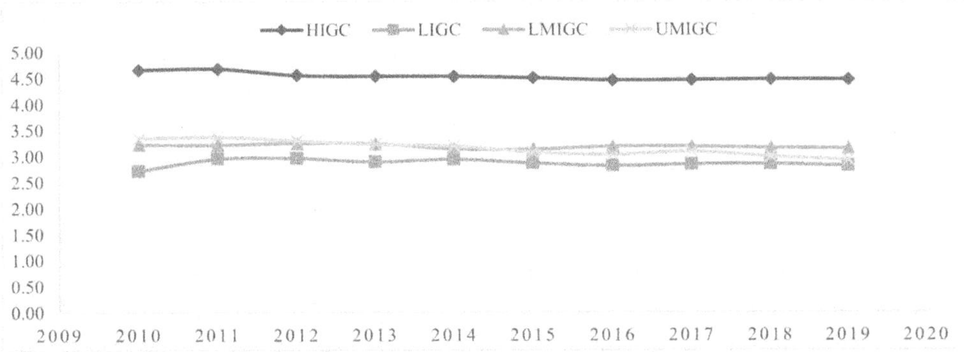

Figure 15. Compulsory education, duration (years)
Source: WDI (World Bank).

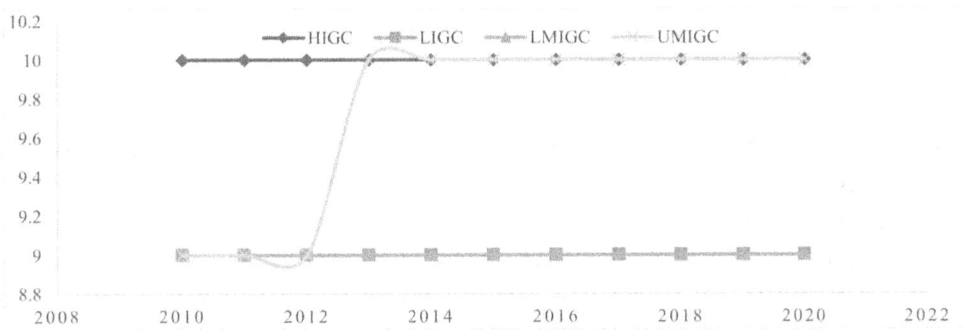

Figure 16. Secondary school enrollment (% gross)
Source: WDI (World Bank).

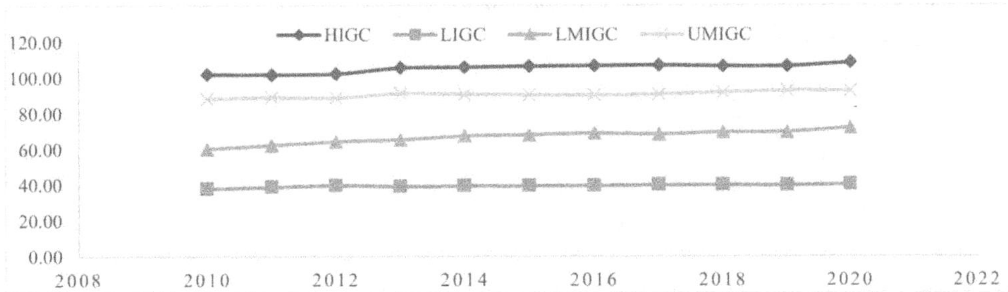

lower-middle and low-income group countries have given crucial attention to improve their position in digitalization. Despite that, LMIGCs and LIGCs have a relatively low position in all indicators of ICTs skills. Hence, it is advisable that LMIGCs and LIGCs should focus their policies to nurture a supportive infrastructure of digitalization. Otherwise, high diversity in digitalization will enhance more inequality across countries.

The progress of digitalization of undertaken countries is measured in term of information and communication technology evolution index (*ICTEI*) that was the composition of ICT use ICT access and ICT skills related 16 variables. Subsequently, overall performance of HIGCs, UMIGCs, LMIGCs and LIGCs is presented in Figure 17. Trend of this figure indicates that HIGCs and UMIGCs have improved their performance in digitalization. While, LMIGCs and LIGCs have an extremely low position in digitalization. It could materialize due to existence of high diversity in indicators which were considered to develop *ICTEI* for different income group countries.

Discussion on the Progress of Digitalization in Each Income Group Countries

The values of *ICTEI* lie between 0.300 – 0527 across HIGCs (Figure 18). The values of ICTEI lie between 0.198 - 0.409 across UMIGCs (Figure 19). The values of ICTDI lie between 0.055 - 0.351 across LMIGCs (Figure 20). The estimates provide a confirmation that there was high diversity in the progress of digitalization across 109 countries. As most HIGCs have a high per capita income, low population growth, low unemployment, better R&D ecosystem and S&T development, better transparency in government policies and financial market, high R&D expenditure, better skilled labour force, better ICTs infrastructure and appropriate job opportunities for working population. Therefore, most countries like Netherlands, South Korea and

Figure 17. Information and communication evolution index (ICTEI)
Source: Author's estimation.

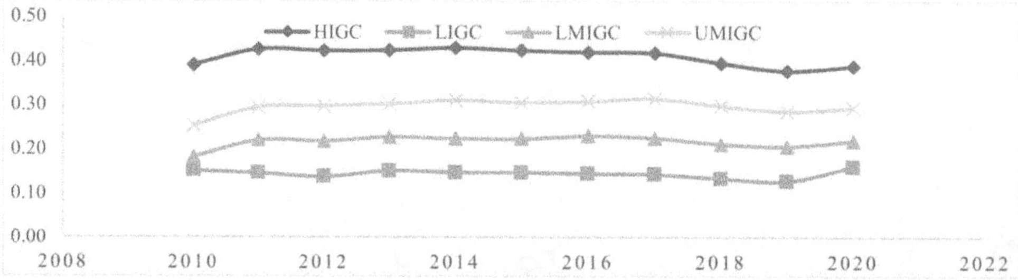

Figure 18. Relative progress of HIGCs in digitalization
Source: Author's estimation.

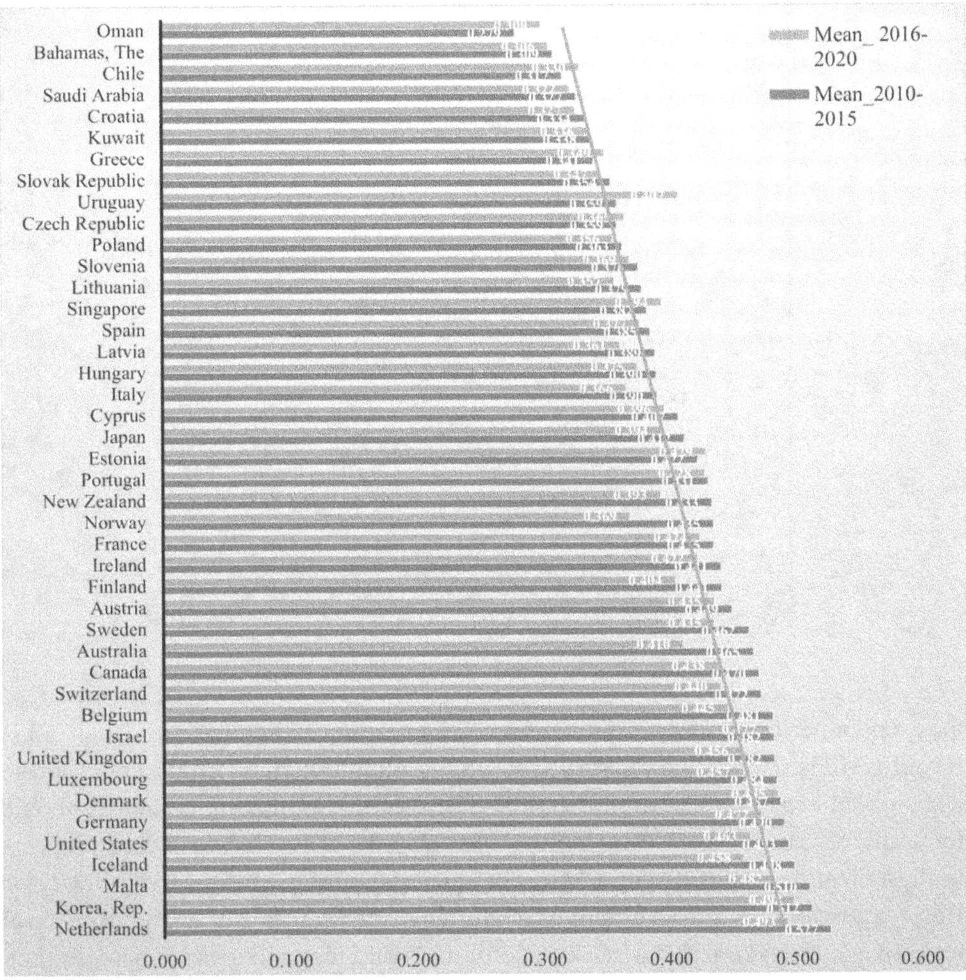

Malta have created a better ecosystem of digitalization. While, Iceland, United States of America, Germany, Denmark, Luxembourg, United Kingdom, Israel, Belgium, Switzerland, Canada, Australia, Sweden, Austria, Finland, Ireland, France, Norway, New Zealand, Portugal, Estonia, Japan and Cyprus also have good performance in digitalization. The values of *ICTDI* were between 0.28 to 0.39 for Italy, Hungry, Latvia, Spain, Singapore, Lithuania, Slovenia, Poland, Czech Republic, Uruguay, Slovak Republic, Greece, Kuwait, Croatia, Saudi Arabia, Chile, Bahamas and Oman. Hence, these countries were in poor position in digitalization.

Costa Rica was in the good position in digitalization in UMIGs. The values of *ICTDI* were between 0.198 to 0.379 for remaining UMIGs. Hence, all UMIGs countries

Figure 19. Relative progress of UMIGCs in digitalization
Source: Author's estimation.

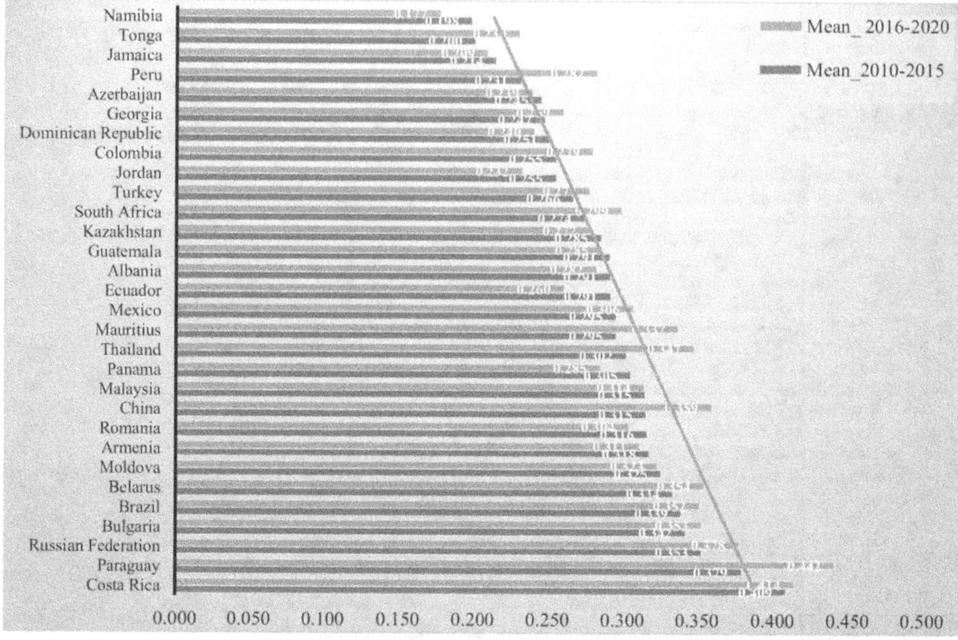

have a poor performance in digitalization. The values of *ICTDI* were less than 0.351 for all LMIGs. Hence, no country from lower-middle income group could create appropriate ecosystem of digitalization. Therefore, low ranking countries should focus all indicators which were included to create ICTEI to increase their progress in digitalization. UMIGIs and LMIGs countries should give extensive priority on R&D expenditure, S&T development, technology transfer and commercialization of technology to develop and discover useful tools and technologies to increase their performance in digitalization. Also, technological skills of people, digital literacy and ICTs infrastructure would be helpful to increase digitalization in UMIGs and LMIGs. The government also should provide fixed telephone, high internet speed, secure internet servers to common citizen in in UMIGIs and LMIGs to improve their performance in digitalization. Singh et al. (2023c) has highlighted that fixed telephone help to increase effective communication across people.

Statistical Properties of Variables

The statistical property of all indicators is presented in Table 4. As per the minimum and maximum values of variables, it is concluded that most variables have a significant variability in HIGCs, UMIGCs and LMIGCs&LIGCs. The statistical

Figure 20. Relative progress of LMIGCs&LIGCs in digitalization
Source: Author's estimation.

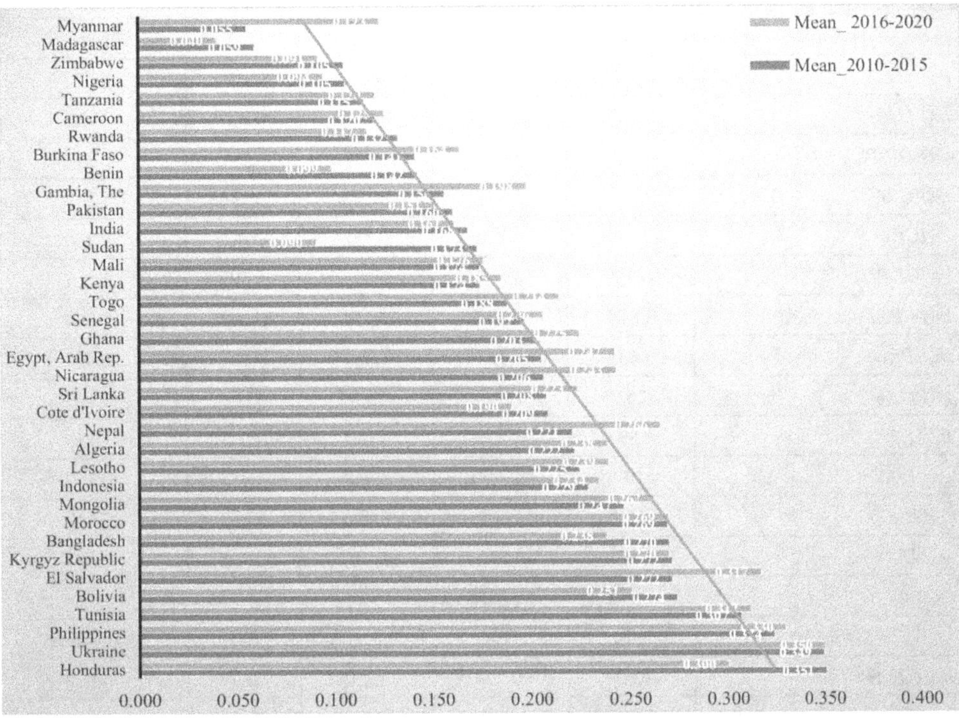

values of standard deviation (SD) were also found nearly 1 for most variables (except, scientific research journal articles) in all groups of countries. The statistical values of skewness were also reported approximately 1 for all variables excluding population growth and access to electricity in all groups of countries, and exports of goods and services in LMIGCs&LIGCs. The estimates, therefore, suggested that most variables were in normal form. However, the estimates also infer that there exists a significant diversity in the statistical properties of undertaken variables across HIGCs, UMIGCs and LMIGCs&LIGCs. Hence, the impact of explanatory variables on digitalization and economic development would be dissimilar in HIGCs, UMIGCs and LMIGCs&LIGCs.

Correlation Coefficients of *ICTEI* and *ED* With Explanatory Variables

The results based on correlation coefficient analytical technique imply that *ICTEI* were positively correlated with economic development, scientific research journal articles, exports of goods and services, foreign direct net investment inflow, access

Table 4. Statistical explanation of dependent and independent variables

HIGCs					
Variables	**Min**	**Max**	**Mean**	**SD**	**Skewness**
ln(ICTEI)	-1.347	-0.615	-0.903	0.154	-0.322
ln(ED)	9.293	11.595	10.328	0.569	0.033
ln(FDINI)	-6.394	5.635	1.093	1.517	-0.318
ln(EGS)	2.316	5.325	3.888	0.586	0.222
ln(ICP)	-4.791	2.278	0.152	1.12	-1.268
ln(IGDPD)	-6.299	3.467	0.374	1.198	-1.267
ln(LFPRT)	3.874	4.353	4.114	0.094	-0.069
ln(EPR)	3.631	4.323	4.034	0.123	-0.467
ln(PGR)	-8.517	1.995	-0.549	1.19	-1.509
ln(AE)	4.597	4.605	4.605	0.001	-7.232
ln(STJA)	1.8	12.979	8.968	1.973	-0.757
ln(GCF)	2.476	4.002	3.096	0.207	-0.213
ln(MER)	-0.984	0.483	-0.197	0.308	-0.181
UMIGCs					
Variables	**Min**	**Max**	**Mean**	**SD**	**Skewness**
ln(ICTEI)	-1.804	-0.754	-1.237	0.2	-0.229
ln(ED)	7.721	9.621	8.76	0.4	-0.049
ln(FDINI)	-3.068	2.539	1.013	0.765	-0.848
ln(EGS)	2.386	4.465	3.491	0.432	-0.033
ln(ICP)	-4.047	4.081	1.153	0.909	-1.181
ln(IGDPD)	-5.828	4.321	1.215	1.088	-1.692
ln(LFPRT)	3.619	4.379	4.09	0.147	-1.05
ln(EPR)	3.421	4.342	4	0.18	-0.765
ln(PGR)	-6.079	1.691	-0.371	1.106	-1.561
ln(AE)	3.745	4.605	4.562	0.129	-4.592
ln(STJA)	-0.777	13.177	6.903	2.631	0.057
ln(GCF)	2.48	3.843	3.166	0.252	0.352
ln(MER)	-1.309	-0.127	-0.748	0.232	0.074
LMIGCs&LIGCs					
Variables	**Min**	**Max**	**Mean**	**SD**	**Skewness**
ln(ICTEI)	-3.468	-0.968	-1.688	0.464	-1.074
ln(ED)	6.092	8.388	7.38	0.616	-0.082

continued on following page

Table 4. Continued

ln(FDINI)	-4.004	3.782	0.731	0.918	-0.265
ln(EGS)	-0.831	4.081	3.136	0.666	-2.822
ln(ICP)	-2.586	6.323	1.362	1.112	-0.224
ln(IGDPD)	-3.416	6.325	1.405	1.199	-0.266
ln(LFPRT)	3.686	4.462	4.119	0.193	-0.018
ln(EPR)	3.552	4.455	4.053	0.222	-0.091
ln(PGR)	-3.214	1.15	0.48	0.631	-1.727
ln(AE)	2.272	4.605	4.103	0.532	-1.214
ln(STJA)	2.344	11.819	5.973	2.032	0.52
ln(GCF)	2.003	4.063	3.17	0.339	-0.34
ln(MER)	-2.142	-0.568	-0.992	0.218	-0.906

Source: Author's estimation.

to electricity and market exchange rate (Table 5). Hence, these variables will be helpful to increase digitalization. Economic development was also positive correlated with scientific research journal articles, information and communication technology evolution index, foreign direct investment net inflows, access to electricity and market exchange rate. Hence, these variables were seemed as vital variables to enhance economic development. While, economic development and digitalization were negatively associated with inflation consumer price and inflation GDP deflator and population growth.

Table 5. Correlation coefficients of ICTEI with economic development and other variables

Variables	*ICTEI*	*ED*	*STJA*	*EGS*	*FDINI*	*ICP*	*IGDPD*	*PGR*	*AE*	*MER*
ICTEI	1									
ED	0.709[b]	1								
STJA	0.264[b]	0.240[b]	1							
EGS	0.456[b]	0.498[b]	-0.139[b]	1						
FDINI	0.176[b]	0.191[b]	-0.061[a]	0.356[b]	1					
ICP	-0.174[b]	-0.110[b]	-0.040	-0.067[b]	-0.027	1				
IGDPD	-0.172[b]	-0.105[b]	-0.048[a]	-0.052[a]	-0.026	0.955[b]	1			
PGR	-0.452[b]	-0.178[b]	-0.198[b]	-0.096[b]	-0.043	0.043	0.060[a]	1		
AE	0.703[b]	0.393[b]	0.162[b]	0.299[b]	0.091[b]	-0.126[b]	-0.121[b]	-0.488[b]	1	
MER	0.719[b]	0.854[b]	0.249[b]	0.248[b]	0.141[b]	-0.136[b]	-0.136[b]	-0.25 [b]	0.360[b]	1

Source: Author's estimation. [b] show that correlation coefficient is significant at the 0.01 level and [a] show that correlation coefficient is significant at the 0.05 level.

Explanation of Regression Results

The regression coefficients of *IV*s with *ED* are presented in Table 6. The coefficients of *IV*s with *ICTEI* are given in Table 7. The coefficient of *ICTEI* with *ED* were reported positive and statistically significant in HIGCs, UMIGCs and LMIGCs&LIGCs. The results infer that economic development is expected to be increased by 1.34%, 0.42% and 0.08% in HIGCs, UMIGCs and LMIGCs&LIGCs, respectively as 1% increase in digitalization. Raeskyesa and Lukas (2019); Habibi and Zabardast (2020); Myovella et al. (2020); Sahoo et al. (2021); Nguyen (2021) have also reported a positive impact of digitalization and ICT indicators on economic development. The coefficient of *ED* with *ICTEI* was seemed positive in HIGCs, UMIGCs and LMIGCs&LIGCs. Hence, economic development increases as digitalization increases. The result also claimed that economic development and digitalization has a positive and causal association with each other's.

Foreign direct investment net inflows assist to increase economic growth and development in multiple ways. It helps to create physical assets, infrastructural

Table 6. Impact of ICTEI and control variables on economic development

Group of Countries	HIGC		UMIGC		LMIGC&LIGC	
Number of obs.	473		330		396	
Number of groups	43		30		36	
R-squared	0.5575		0.3874		0.6166	
Wald Chi²(10)	1752.58		9528.12		7400.58	
Prob > Chi²	0.000		0.000		0.000	
ln(ED) = DV	*Reg. Coef.*	*Std.*	*Reg. Coef.*	*Std.*	*Reg. Coef.*	*Std.*
ln(ICTEI)	1.3462*	0.1339	0.4280*	0.0558	0.0758*	0.0349
ln(FDINI)	0.0219*	0.0098	0.0231	0.0194	-0.0120	0.0233
ln(EGS)	0.0513*	0.0246	0.0669*	0.0134	0.1321*	0.0253
ln(ICP)	-0.0346***	0.0181	-0.0241	0.0201	-0.0005	0.0109
ln(LFPRT)	1.4121*	0.5437	-0.6289*	0.1337	0.5148***	0.2957
ln(EPR)	0.7156	0.5277	1.03634*	0.0988	-0.8292*	0.2338
ln(PGR)	0.1101*	0.0167	0.0733*	0.0166	-0.0237	0.0278
ln(AE)	83.9914*	23.6317	0.1344*	0.0573	0.6573*	0.0550
ln(STJA)	0.0848*	0.0090	0.0563*	0.0028	0.0460*	0.0046
ln(GCF)	0.0100	0.0679	0.1318*	0.0386	-0.0557***	0.0315
Con. Coef.	-384.8856*	108.7900	7.3223*	0.4369	5.5592*	0.5415

Source: Author's estimation. *, ** and *** show that coefficients are statistically significant at 1%, 5% and 10% significance level, respectively.

Table 7. Impact of economic development and control variables on ICTEI

Group of Countries	HIGC		UMIGC		LMIGC&LIGC	
Number of Obs.	473		330		396	
Number of groups	43		30		36	
R-squared	0.585		0.3536		0.5455	
Wald $Chi^2(10)$	2437.72		1113.63		824.84	
$Prob > Chi^2$	0.000		0.000		0.000	
ln(ICTEI) = DV	*Reg. Coef.*	*Std.*	*Reg. Coef.*	*Std.*	*Reg. Coef.*	*Std.*
ln(ED)	0.0699*	0.0265	0.1037*	0.0216	0.0068	0.0292
ln(FDINI)	0.0119*	0.0036	-0.0223***	0.0124	-0.0252	0.0177
ln(EGS)	0.1028*	0.0119	0.0547*	0.0198	0.1715*	0.0244
ln(IGDPD)	0.0085***	0.0051	-0.0085	0.0087	-0.0035	0.0139
ln(EPR)	0.1522*	0.0272	0.0551***	0.0325	-0.5158*	0.0599
ln(PGR)	0.0029	0.0037	-0.0305*	0.0065	-0.0990*	0.0200
ln(AE)	14.7969**	6.6609	0.4864*	0.0552	0.4096*	0.0552
ln(STJA)	0.0365*	0.0020	0.0138*	0.0019	-0.0297*	0.0064
ln(MER)	0.3941*	0.0534	0.0052	0.0586	0.2821*	0.0671
Con. Coef.	-69.5991**	30.6689	-4.8450*	0.2265	-1.3380*	0.2935

Source: Author's estimation. *, ** and *** show that coefficients are statistically significant at 1%, 5% and 10% significance level, respectively.

development and employment opportunities. It also maintains the money flow and financial stability in money market. As a result, it will be positive to increase economic growth and development. The coefficient of *FDINI* with economic development was found positive in HIGCs and UMIGCs. Hence, the estimated revealed that economic development may be increased by 0.02% and 0.23% in HIGCs and UMIGCs, respectively as foreign direct investment net inflows increase by 1%. The mentioned relation is consistent with previous studies such as Sahoo et al. (2021); Soomro et al. (2022). In contrary, the impact of foreign direct investment net inflows on economic development was seemed negative in LMIGC&LIGC. Soomro et al. (2022) have also noticed a negative impact of FDI on economic growth in BRICS countries. Foreign direct investment net inflows showed a negative impact on digitalization in UMIGCs and LMIGCs&LIGCs. While, foreign direct investment has a positive impact on digitalization in HIGCs. Hence, impact of foreign net investment net inflows on digitalization were varied in the different income group countries. Most upper-middle and lower-middle & low-income group countries could not create a favourable ecosystem to use foreign direct investment for creation of physical assets and infrastructural development. Hence, foreign direct investment forced a negative impact on economic development and digitalization in UMIGCs and LMIGCs&LIGCs.

Exports of goods and services is an effective indicator which measures the impact of trade openness on economic activities. Trade openness is a key player to enhance trade relationship across countries. Hence, it is expected that trade openness helps to promote economic growth and development (Nguyen, 2021; Soomro et al., 2022). The coefficient of exports of goods and services with economic development was appeared positive and statistically significant. The estimates claimed that economic development to be increased by 0.05%, 0.06% and 0.13% in HIGCs, UMIGCs and LMIGCs&LIGCs, correspondingly due to 1% increase in exports of goods and services. The similar relationship was also observed by Singh and Ashraf (2020); Nguyen (2021). The positive coefficient of *EGS* with *ICTEI* infer that process of digitalization also improves as exports of goods and services increases in HIGC, UMIGC and LMIGC&LIGC. The results claimed that digitalization is likely to be increased by 10%, 0.05% and 0.17% in HIGC, UMIGC and LMIGC&LIGC, respectively as exports of goods and service increase by 1%.

The coefficient of inflation consumer price with economic development was negative in HIGCs, UMIGCs and LMIGCs&LIGCs. The estimate reveals that economic development decline as inflation increases. The negative relationship of inflation with economic development was also observed by Tiutiunyk et al. (2021); Singh and Kumar (2022a); Singh et al. (2023a). The coefficient of inflation GDP deflator with *ICTEI* was positive in HIGCs, and negative in UMIGCs and LMIGCs&LIGCs. Hence, the estimates indicates that inflation have a diverse impact on digitalization in HIGCs, UMIGCs and LMIGCs&LIGCs. High income countries have higher per capita income as compared to UMIGCs and LMIGCs&LIGCs. Therefore, inflation have a minimum adverse impact on purchasing power of people in HIGCs. While, inflation has a significant and negative impact on purchasing power and economic capacity of people in UMIGCs and LMIGCs&LIGCs. Inflation, therefore, showed a positive impact on digitalization in HIGCs, and negative impact on digitalization in UMIGCs and LMIGCs&LIGCs.

Labour force participation rate has a positive impact on economic development in HIGCs and LMIGCs&LIGCs. While, its impact was reported negative in UMIGCs. The estimates reveal that economic development to be increased by 0.14% as labour force participation rate increases by 1% in HIGCs. Further, it seemed that labour force participation rate has a diverse impact on economic development due to engagement of working population in different sector of HIGCs, UMIGCs and LMIGCs&LIGCs.

Employment to population ratio shows the number of people employed in total working-age population. It also measures that effectiveness of a country to provide jobs for working population. Hence, employment to population ratio produce a positive impact on economic development. Accordingly, the impact of employment to population ratio on economic development was positive in HIGCs and UMIGCs. The results imply that economic development to be increased by 0.71% and 1.03% in HIGCs and UMIGCs, respectively as employment to population ratio increases by 1%. The impact of employment to population ratio on economic development was reported negative in LMIGCs&LIGCs. Employment to population ratio increased due to rising of vulnerable employment in LMIGCs&LIGCs. While, vulnerable employment is negatively associated with economic development (Aly, 2020; Singh et al., 2020a). Hence, employment to population ratio produce a negative impact on economic development in LMIGCs&LIGCs. The coefficient of *EPR* with *ICTEI* was found positive in HIGCs and UMIGCs. Hence, digitalization will be beneficial due to increase in employment to population ratio in HIGCs and UMIGCs. The estimates revel that digitalization to be increased by 0.15% and 0.05% in HIGCs and UMIGCs, respectively as employment to population ratio increases by 1%. While, employment to population ratio enforce a negative impact on digitalization due to presence of vulnerable employment in LMIGCs&LIGCs.

The coefficient of *PGR* with *ED* was appeared positive in HIGCs and UMIGCs. Hence, the estimates imply that population growth has a positive impact on economic development in HIGCs and UMIGCs. Economic development is projected to be increased by 0.11% and 0.07% in HIGCs and UMIGCs, respectively as population growth increase by 1%. Previous studies also argued that population growth helps to create demand of goods and services, meet the demand of labour in labour marker, and increase industrialization and urbanization (Singh et al., 2020a; Singh & Kumar, 2022a; Singh et al., 2023a). Therefore, population growth may be positive to increase economic development in HIGCs and UMIGCs. While, LMIGCs&LIGCs have high population growth with large size. Henceforth, population growth produced a negative impact on economic development in these countries. The coefficient of *PGR* with *ICTEI* was seemed positive in HIGCs, and negative in UMIGCs and LMIGCs&LIGCs. The estimates, therefore, infer that population growth will be positive to improve digitalization in HIGCs. While, high population growth will create a barrier in the progress of digitalization in UMIGCs and LMIGCs&LIGCs.

Most production activities in manufacturing sector and economic activity of individual person cannot be achieved without electricity. It is a best indicator of infrastructural development. Hence, access to electricity is supportive to enhance social and economic development in multiple ways (Sahoo et al., 2021; Jyoti & Singh, 2023). Furthermore, availability of electricity is a crucial driver to increase digitalization. Therefore, the coefficient of *AE* with *ED* and *ICTEI* were observed positive and statistically significant in all income group countries. The estimates imply that economic development to be improved by 83.99%, 0.13% and 0.65% in HIGCs, UMIGCs and LMIGCs&LIGCs, respectively as access to electricity increases by 1%. The digitalization would be increased by 14.79%, 0.48% and 0.41% in HIGCs, UMIGCs and LMIGCs&LIGCs, respectively due to 1% increase in accessibility of electricity.

Science & technological development (S&TD) brings several positive alternatives to increase social-economic development (Singh et al., 2020a; Singh et al., 2019c). Production, efficiency and productivity of manufacturing, service and agricultural sectors could be improved due to S&TD (Singh et al., 2019b; Ashraf and Singh, 2022). Accordingly, growth of most sector increased due to application on S&TD in production activities (Kumar et al., 2017; Ashraf and Singh, 2022). S&TD also helps to discover innovation, process, knowledge and idea which can be used to develop new goods and services in the product market (Singh et al., 2020a). S&TD, therefore, helps to create innovation in the market, business and jobs. Furthermore, S&TD is an originator of many technologies which enhance digitalization and develop a digital platform. In this study, scientific and technical journal articles (*STJA*) were used to capture the impact of S&TD on economic development and digitalization (Singh et al., 2020a). The regression coefficient of *STJA* with *ED* was detected positive and statistically significant in HIGCs, UMIGCs and LMIGCs&LIGCs. The estimates also imply that economic development is expected to be increased by 0.08%, 0.05% and 0.04% in HIGCs, UMIGCs and LMIGCs&LIGCs, respectively due to 1% increase in S&TD. The impact of S&TD on digitalization was observed positive in HIGCs and UMIGCs. While, S&TD showed a negative impact on digitalization in LMIGCs&LIGCs. LMIGCs&LIGCs were seemed unable to apply S&TD in production activities efficiently due to low skills of people and low technological absorption capacity of manufacturing firms (Singh et al., 2019b). Also, imitation rate of technology is higher in these countries due to ineffective mechanism of intellectual property rights regime (Singh et al., 2020a). Hence, the impact of S&TD on digitalization may be negative in LMIGCs&LIGCs.

Gross capital formation helps to create physical assets in terms of land improvement, plants, machine, equipment, road construction, building construction, school, hospitals and etc. Hence, gross capital formation would be positive to increase economic development (Sahoo et al., 2021). The results indicate that gross capital formation have a positive impact on economic development in HIGCs and UMIGCs. While, it showed a negative impact on economic development in LMIGCs&LIGCs due to extensive pressure of population on available resources. The coefficient of *MER* with *ICTEI* were reported positive in all income group countries. Hence, the estimates provide a support that digitalization depend on market exchange rate that reflect the foreign trade across countries. Export and imports of ICTs, digitalization enhancing products and technological development are also determined by market exchange rate.

CONCLUSION AND POLICY GUIDELINE

The descriptive results of this chapter showed the progress of most indicators of ICT access, ICT use and ICT skills have increased consistently in HIGCs during 2010-2022. Thus, HIGCs were in better position in digitalization. Also, UMIGCs have an increasing trend in most indicators of ICT which were positive to increase digitalization. Most indicators of ICT have a positive trend in LMIGCs and LIGCs. Despite that, LMIGCs and LIGCs have relatively lower position in all indicators of ICTs. Therefore, LMIGCs and LIGCs should implement effective policies to nurture an appropriate infrastructure of digitalization.

The integrated and relative progress of digitalization for 109 countries was measured as information and communication technology evolution index (*ICTEI*) which was the composition of 16 indicators in the domain of ICT use ICT access and ICT skills related variables. HIGCs and UMIGCs could improve their performance in digitalization due to their better position in digitalization supporting variables. While, LMIGCs&LIGCs have a lower position in digitalization. It is happened due to existence of high diversity in indicators which were used to develop *ICTEI* across different income group countries. Thereupon, HIGCs, UMIGCs and LMIGCs&LIGCs have a diversity in digitalization within group of countries.

The results based on correlation coefficient analytical technique imply that digitalization was positively associated with economic development, scientific and technical research journal articles, exports of goods and services, foreign direct net investment inflows, access to electricity and market exchange rate. Economic development was positively correlated with scientific research journal articles, digitalization, foreign direct investment net inflows, access to electricity and market exchange rate. Economic development and digitalization were also negatively associated with inflation consumer price, inflation GDP deflator and population growth.

The regression coefficient of *ICTEI*, exports of goods and services, and scientific and technical journal articles with *ED* were appeared positive in HIGCs, UMIGCs, and LMIGCs&LIGCs. Hence, the estimates infer that digitalization, trade openness and S&TD have a positive impact on economic development. While, economic development was negatively impacted due to increase in inflation consumer price in HIGCs, UMIGCs, and LMIGCs&LIGCs. The impact of foreign direct investment net inflows, labour force participation rate, population growth, access to electricity and gross capital formation on economic development were differed in HIGCs, UMIGCs, and LMIGCs&LIGCs. Economic development, exports of goods and services, access to electricity and market exchange rate produced a positive impact on digitalization in HIGCs, UMIGCs, LMIGCs&LIGCs. Foreign direct investment net inflows, inflation GDP deflator, employment to population ratio, population growth, S&TD have diverse impact on digitalization in HIGCs, UMIGCs and LMIGCs&LIGCs.

The estimates provide evidence that economic development and digitalization have a positive and causal relationship. Hence, all income group countries should give significant priority to economic development and digitalization promoting factors. However, the impact of explanatory variables on economic development and digitalization were not similar across different income group countries. Hence, following suggestions can be conducive to increase economic development and digitalization in HIGCs, UMIGCs and LMIGCs&LIGCs: HIGCs should increase exports of goods and services, foreign direct investment, population growth and market exchange rate, employment to population ratio and S&TD to maintain digitalization and economic development. For UMIGCs, exports of goods and services, employment to population ratio, S&TD and gross capital formation will be positive to increase economic development and digitalization. UMIGCs and LMIGCs&LIGCs should control population growth and inflation to achieve a significant benefit from digitalization for further sustainable economic development. LMIGCs&LIGCs have a poor position in digitalization and economic development. Hence, these countries have several challenges to improve their position in digitalization and economic development. LMIGCs&LIGCs should increase investment in ICT infrastructure (Myovella et al., 2020; Özsoy et al., 2022; Aksentijevic et al., 2021). Extensive R&D expenditure, appropriate R&D ecosystem in research organization, science & technological development, human skills, technology transfer and commercialization, infrastructural development will be useful to develop ICTs ecosystem, digital technology, other technologies and digitalization in low-income group countries (Sahoo et al., 2021). Better education and training facilities for youth generation will increase their capacity to use ICT indicators in production activities in developing countries (Raeskyesa & Lukas, 2019). However, the chapter could not assess the influence of R&D expenditure, human skills, technology transfer and commercialization and infrastructural development on digitalization. Therefore, further study can include above-mentioned indicators in empirical analysis to validate the findings of this chapter.

This chapter has following contribution in the theoretical and empirical literature:

- It explained the progress of digitalization enhancing variables in HIGCs, UMIGCs, LMIGCs and LIGCs using a trend analysis.
- It developed *ICTEI* as integration of 16 different indicators related to ICT access, ICT use and ICT skills across 109 countries during 2010 – 2020 using Z -score method and principal component analysis.
- It described the relative progress of across 109 countries in digitalization as per the estimated values of *ICTEI*.

- It estimated the correlation coefficients of *ICTEI* and economic development with explanatory variables.
- It examined the impact of digitalization on economic development and vice-versa in HIGCs, LMIGCs and LMIGCs&LIGCs using a Cobb-Douglas production function model.
- It suggested several policy proposals to increase the digitalization and economic development in HIGCs, UMIGCs, LMIGCs and LIGCs.
- It also provided a research direction for further study to check the validity of empirical findings of this chapter.

REFERENCES

Afonasova, M. A., Panfilova, E. E., Galichkina, M. A., & Ślusarczyk, B. (2019). Digitalization in economy and innovation: The effect on social and economic process. *Polish Journal of Management Studies*, *19*(2), 22–32. doi:10.17512/pjms.2019.19.2.02

Aksentijevic, N. K., Ježic, Z., & Zaninovic, P. A. (2021). The effects of information and communication technology (ICT) use on human development – A macroeconomic approach. *Economics*, *9*(128), 1–12.

Aleksandrova, A., Truntsevsky, Y., & Polutova, M. (2022). Digitalization and its impact on economic growth. *Revista de Economia Política*, *42*(2), 424–441. doi:10.1590/0101-31572022-3306

Ali, M. A., Alam, K., & Taylor, B. (2019). Incorporating affordability, efficiency, and quality in the ICT development index: Implications for index building and ICT policymaking. *The Information Society*, *36*(2), 71–96.

Aly, H. (2020). Digital transformation, development and productivity in developing countries: Is artificial intelligence a cure or a blessing? *Review of Economics and Political Science*, *7*(4), 1–19.

Appiah-Otoo, I., & Song, N. (2021). The impact of ICT on economic growth-comparing rich and poor countries. *Telecommunications Policy*, *45*(2), 1–15. doi:10.1016/j.telpol.2020.102082

Arabi, A. M. K., & Allah, B. A. W. (2017). The impact of information and communication technology (ICT) development on economic growth in Sudan: An application of ARDL bounds testing approach. *Archives of Business Research*, *5*(3), 155–165.

Ashraf, S. N., & Singh, A. K. (2019). Does entrepreneurship ecosystem have a long-term relationship with economic growth in selected economies? A statistical investigation. In Misra S., Shukla S. and Batthini G. (Eds.), *Proceedings of the 13th Biennial Conference on Entrepreneurship, EDII Ahmedabad (Vol. 1*, pp. 176-187). Bookwell Publishing House, New Delhi.

Ashraf, S. N., & Singh, A. K. (2022). Implications of appropriate technology and farm inputs in the agricultural sector of Gujarat: Empirical analysis based on primary data. *Agricultural Economics Research Review*, *35*(2), 59–77. doi:10.5958/0974-0279.2022.00031.3

Bakari, S. (2022). The impact of digitalization and patent on economic growth in Romania. *MPRA Paper No. 114259.* MPRA. https://mpra.ub.uni-muenchen.de/114259/

Bakry, W., Nghiem, X. H., Farouk, S., & Vo, X. V. (2023). Does it hurt or help? revisiting the effects of ICT on economic growth and energy consumption: A nonlinear panel ARDL approach. *Economic Analysis and Policy*, *78*(1), 597–617. doi:10.1016/j.eap.2023.03.026

Benjamin, B., & Woerter, M. (2019). Is this time different? How digitalization influences job creation and destruction. *Research Policy*, *48*(8), 1–10.

Carbo-Valverde, S. (2017). The impact on digitalization on banking and financial stability. *Journal of Financial Management. Markets and Institutions*, *5*(1), 133–140.

Dahlman, C., Mealy, S., & Wermelinger, M. (2016). *Harnessing the digital economy developing countries. OECD Development Centre,* (Working Paper No. 334). OECD Publishing, Paris.

Dahmani, M., Mabrouki, M., & Youssef, A. B. (2022). ICT, trade openness and economic growth in Tunisia: What is going wrong? *Economic Change and Restructuring*, *55*(1), 2317–2336. doi:10.100710644-022-09388-2

Dobrota, M., Jeremic, V., & Markovik, A. (2012). A new perspective of the ICT development index. *Information Development*, *28*(4), 271–280. doi:10.1177/0266666912446497

Farhadi, M., Ismail, R., & Fooladi, M. (2012). Information and communication technology use and economic growth. *PLoS One*, *7*(11), e48903. doi:10.1371/journal.pone.0048903 PMID:23152817

Gerpott, T. J., & Ahmadi, N. (2015). Composite indices for the evaluation of a country's information technology development level: Extension of the IDI of the ITU. *Technological Forecasting and Social Change*, *98*(1), 174–185. doi:10.1016/j.techfore.2015.03.012

Habibi, F., & Zabardast, M. A. (2020). Digitalization, education and economic growth: A comparative analysis of Middle East and OECD countries. *Technology in Society*, *63*(10), 23–47. doi:10.1016/j.techsoc.2020.101370

Hernandez, K., Faith, B., Martín, P. P., & Ramalingam, B. (2016). *The impact of digital technology on economic growth and productivity, and its implications for employment and equality: An evidence review. Evidence Report No. 207.* Institute of Development Studies., https://www.ids.ac.uk/publications/the-impact-of-digital-technology-on-economic-growth-and-productivity-and-its-implications-for-employment-and-equality-an-evidence-review/

Irtyshcheva, I., Stehnei, M., Popadynets, N., Bogatyrev, K., Boiko, Y., Kramarenko, I., Senkevich, O., Hryshyna, N., Kozak, I., & Ishchenko, O. (2021). The effect of digital technology development on economic growth. *International Journal of Data and Network Science*, *5*(1), 25–36. doi:10.5267/j.ijdns.2020.11.006

Jio, S., & Sun, Q. (2021). Digital economic development and its impact on economic growth in China: Research based on the perspective of sustainability. *Sustainability (Basel)*, *13*(1), 1–18. doi:10.3390u131810245

Jyoti, B., & Singh, A. K. (2023). Does digitalization have a causal relationship with economic development? An experience from a country-wise panel data statistical analysis. In R. S. Nazlı & G. Sari (Eds.), *Perspectives on Society and Technology Addiction*. IGI Global. doi:10.4018/978-1-6684-8397-8.ch003

Kumar, A., Ahmad, M. M., & Sharma, P. (2015). Carbon emission and global food security: A cross country analysis. *PENCIL Publication of Agricultural Sciences*, *2*(1), 7–24.

Kumar, A., Ahmad, M. M., & Sharma, P. (2017). Influence of climatic and non-climatic factors on sustainable food security in India: A statistical investigation. *International Journal of Sustainable Agricultural Management and Informatics*, *3*(1), 1–30. doi:10.1504/IJSAMI.2017.082917

Kumar, A., & Sharma, P. (2013). Impact of climate variation on agricultural productivity and food security in rural India. *Economics Discussion Papers, No. 2013-43*. Kiel Institute for the World Economy Germany. http://www.economics-ejournal.org/economics/discussionpapers/2013-43

Kutsuri, G. N., Kamberdieva, S. S., Dedegkaev, V. K. H., Sopoeva, I. A., & Shelkunova, T. G. (2019). Impact of digitalization on improvement of economy, IT and internet of business. *Journal of Physics: Conference Series*, *1399*(3), 1–6. doi:10.1088/1742-6596/1399/3/033008

Li, R., & Gospodarik, C. G. (2021). The impact of digital economy on economic growth based on pearson correlation test analysis. In: J. Jansen, B., Liang, H., Ye, J. (eds) *International Conference on Cognitive based Information Processing and Applications (CIPA 2021). Lecture Notes on Data Engineering and Communications Technologies, vol 85*. Springer, Singapore. 10.1007/978-981-16-5854-9_3

Ma, D., & Zhu, Q. (2022). Innovation on emerging economies: Research on the digital economy driving high quality green development. *Journal of Business Research, 145*(1), 891–813. doi:10.1016/j.jbusres.2022.03.041

Minges, M. (2015). *Exploring the relationship between broadband and economic growth*. Background Paper, World Development Report 2016.

Myovella, G., Karacuka, M., & Haucap, J. (2020). Digitalization and economic growth: A comparative analysis of Sub-Saharan Africa and OECD economies. *Telecommunication Policy, 44*(2).

Nguyen, V. B. (2021). The digitalization – economic growth relationship in developing countries and the role of governance. *Scientific Annals of Economics and Business, 68*(4), 481–493. doi:10.47743aeb-2021-0028

Ogloblina, E. V., Seredina, M. I., Altunina, J. O., Kodolov, V. A., & Lebedev, K. A. (2020). Socio-economic consequences of digital development of the economy. *Revista Inclusiones*, *7*(1), 421–430.

Özsoy, S., Ergüzel, O. Ş., Ersoy, A. Y., & Saygılı, M. (2022). The impact of digitalization on export of high technology products: A panel data approach. *The Journal of International Trade & Economic Development*, *31*(2), 277–298. doi:10.1080/09638199.2021.1965645

Pradhan, R. P., Sarangi, A. K., Maity, C., & Behera, R. R. (2022). ICT infrastructure and economic growth in G-20 Countries: New insights form ARDL modelling. *Journal of Developing Areas*, *56*(1), 47–58. doi:10.1353/jda.2022.0009

Raeskyesa, D. G. S., & Lukas, E. N. (2019). Does digitalization increase economic growth? Evidence from ASEAN8 countries. *Jurnal Ekonomi Indonesia*, *8*(2), 267–278. doi:10.52813/jei.v8i2.33

Rath, B. N., & Hermawan, D. (2019). Do information and communication technologies foster economic growth in Indonesia? *Bulletin of Monetary Economics and Banking*, *22*(1), 103–122. doi:10.21098/bemp.v22i1.1041

Sahoo, D., Nayak, S., & Behera, J. (2021). *Digitalization and Economic Performance of Two Fast-Growing Asian Economies: India and the People's Republic of China. ADBI Working Paper 1243*. Asian Development Bank Institute. Available https://www.adb.org/publications/digitalization-economic-performance-fast-growing-asian-economies-india-prc

Sharma, P., & Singh, A. K. (2017). Association of state-wise food security index with climatic factors in India: Evidence from state-wise panel data. *Journal of Global Agriculture and Ecology*, *6*(3), 196–205.

Singh, A. K., Ahmad, M. M., & Sharma, P. (2017). Implications of socioeconomic factors on food security in selected economies: An empirical assessment. *Journal of Global Economics. Management and Business Research*, *8*(2), 103–115.

Singh, A. K., & Ashraf, S. N. (2020). Association of entrepreneurship ecosystem with economic growth in selected countries: An empirical exploration. *Journal of Entrepreneurship. Business Economics (Cleveland, Ohio)*, *8*(2), 36–92.

Singh, A. K., Ashraf, S. N., & Arya, A. (2019b). Estimating factors affecting technical efficiency in Indian manufacturing sector. *Eurasian Journal of Business and Economics*, *12*(24), 65–86. doi:10.17015/ejbe.2019.024.04

Singh, A. K., Ashraf, S. N., & Ashish, A. (2019c). A conceptual review on economic, business, intellectual property rights and science & technology related activities in Asian economies. [JJEM]. *JNNCE Journal of Engineering & Management*, *3*(2), 1–22.

Singh, A. K., Ashraf, S. N., Sharma, S. K., & Kumar, S. (2023e). Factors affecting farm income and climate adaptation strategies in the farming sector of Gujarat state of India. In B. J. Sadhik, J. Adewole, A. Yenèneh, & L. Abidoye (Eds.), *Opportunities and Challenges in Climate-Friendly Clean Water and Energy Technologies* (pp. 122–147). IGI Global. doi:10.4018/978-1-6684-7303-0.ch006

Singh, A. K., & Issac, J. (2018). Impact of climatic and non-climatic factors on sustainable livelihood security in Gujarat state of India: A statistical exploration. *Agriculture and Food Sciences Research*, *5*(1), 30–46. doi:10.20448/journal.512.2018.51.30.46

Singh, A. K., Issac, J., & Narayanan, K. G. S. (2019a). Measurement of environmental sustainability index and its association with socio-economic indicators in selected Asian economies: An empirical investigation. *International Journal of Environment and Sustainable Development*, *18*(1), 57–100. doi:10.1504/IJESD.2019.098641

Singh, A. K., & Jyoti, B. (2020). Factors affecting firm's annual turnover in selected manufacturing industries of India: An empirical study. *Business Perspectives and Research*, *2*(3), 33–59. doi:10.38157/business-perspective-review.v2i3.206

Singh, A. K., Jyoti, B., Kumar, S., & Lenka, S. K. (2021). Assessment of global sustainable development, environmental sustainability, economic development and social development index in selected economies. *International Journal of Sustainable Development and Planning*, *16*(1), 123–138. doi:10.18280/ijsdp.160113

Singh, A. K., & Kumar, S. (2021). Assessing the performance and factors affecting industrial development in Indian states: An empirical analysis. *Journal of Social Economics Research*, *8*(2), 135–154. doi:10.18488/journal.35.2021.82.135.154

Singh, A. K., & Kumar, S. (2022a). Factors influencing entrepreneurship ecosystem and economic development: Evidence from cross-sectional country-wise imbalanced panel data analysis. *The IUP Journal of Entrepreneurship Development*, *19*(3), 24–47.

Singh, A. K., & Kumar, S. (2022b). Exploring the impact of sustainable development on social-economic, and science and technological development in selected countries: A panel data analysis. *Society & Sustainability*, *4*(1), 55–83. doi:10.38157s.v4i1.405

Singh, A. K., Kumar, S., & Jyoti, B. (2022a). Influence of climate change on agricultural sustainability in India: A state-wise panel data analysis. *Asian Journal of Agriculture*, *6*(1), 15–27. doi:10.13057/asianjagric/g060103

Singh, A. K., Kumar, S., & Jyoti, B. (2022c). Association of sustainable food security with climatic and non-climatic factors in Gujarat state of India: A district-wise panel data investigation. In: Leal Filho, W., Kovaleva, M., Popkova, E. (eds.), Sustainable Agriculture and Food Security. World Sustainability Series. Springer Cham. Springer Nature Switzerland. doi:10.1007/978-3-030-98617-9_16

Singh, A. K., Kumar, S., & Jyoti, B. (2023a). Does entrepreneurship ecosystem have a causal relationship with economic development? A country-wise imbalanced panel data investigation. In H. Magd, S. A. Khan, N. D. Nte, & M. Sarim (Eds.), *Entrepreneurship and Economic Sustainability*. Prestige Publishers.

Singh, A. K., Kumar, S., & Sharma, A. K. (2023c). Does green entrepreneurship ecosystem have causal association with economic development? evidence from country-wise panel data investigation. *Journal of International Business Research*, *22*(2), 1–18.

Singh, A. K., Kumar, S., Sharma, A. K., & Sinha, S. (2022b). Does green entrepreneurship have an association with sustainable development and its components? Evidence from a country-wise panel data investigation. In M. Magd, D. Singh, D. Spicer, & R. T. Syed (Eds.), *International Perspectives on Value Creation and Sustainability Through Social Entrepreneurship*. IGI Global USA. doi:10.4018/978-1-6684-4666-9.ch008

Singh, A. K., Sharma, A. K., & Jyoti, B. (2023d). Does economic development have a causal relationship with environmental degradation? Experience from different income group countries. In G. Sart (Ed.), *Considerations on Education for Economic, Social, and Environmental Sustainability* (pp. 300–333). IGI Global. doi:10.4018/978-1-6684-8356-5.ch015

Singh, A. K., Sharma, S. K., & Lenka, S. K. (2023b). Causality between green entrepreneurship and sustainable development: A cross-country analysis using ARDL model. *The IUP Journal of Applied Economics*, *22*(1), 1–34.

Singh, A. K., Singh, B., & Ashraf, S. N. (2020a). Implications of intellectual property protection, and science and technological development in the manufacturing sector in selected economies. *Journal of Advocacy. Research in Education*, *7*(1), 16–35.

Singh, A. K., Singh, B. J., & Negi, V. (2020b). Does sustainable development have a causal relationship with environmental development? Evidence from a country-wise panel data analysis. *International Journal of Technology Management & Sustainable Development*, *19*(2), 147–171. doi:10.1386/tmsd_00020_1

Solomon, E. M., & van Klyton, A. (2020). The impact of digital technology usage on economic growth in Africa. *Utilities Policy*, *67*(10), 1–12. doi:10.1016/j.jup.2020.101104 PMID:32904493

Soomro, A. N., Kumar, J., & Kumari, J. (2022). The dynamic relationship between FDI, ICT, trade openness, and economic growth: Evidence from BRICS countries. *Journal of Asian Finance. Economics and Business*, *19*(2), 295–303.

Tiutiunyk, I., Drabek, J., Antoniuk, N., Navickas, V., & Rubanov, P. (2021). The impact of digital transformation on macroeconomic stability: Evidence from EU countries. *Journal of International Students*, *14*(3), 220–234. doi:10.14254/2071-8330.2021/14-3/14

Verma, A., Dandgawhal, P. S., & Giri, A. K. (2023). Impact of ICT diffusion and financial development on economic growth in developing countries. *Journal of Economics, Finance and Administrative Service*. doi:10.1108/JEFAS-09-2021-0185

Zhang, Zhao, S., Wan, X., & Yao, Y. (2021). Zhang, W., Zhao, S., Wan, X, & Yao, Y. (2021). Study on the effect of digital economy on high-quality economic development in China. *PLoS One*, *16*(9), e0257365. doi:10.1371/journal.pone.0257365 PMID:34547019

Chapter 4
Machine Learning-Enhanced Text Mining as a Support Tool for Research on Climate Change:
Theoretical and Technical Considerations

Nikodem Rybak
https://orcid.org/0000-0002-7708-253X
The University of Queensland, Australia

Maureen Hassall
https://orcid.org/0000-0003-1191-5778
The University of Queensland, Australia

ABSTRACT

In this chapter, the authors explore the theoretical and practical aspects of using text mining approaches supported by machine learning for the automatic interpretation of bulk literature on a contemporary issue—that of climate change risk analysis. The strengths, weaknesses, and opportunities associated with these approaches are investigated. Text mining provides a way to automate and enhance the analysis of text data. However, contrary to popular belief, text mining analysis is not a completely automated process. As with computer-assisted (or -aided) qualitative data analysis software (CAQDAS), it is an iterative method requiring input from a researcher with expert knowledge and a deliberate approach to the analysis. Given the heterogeneity that generally characterizes climate disclosures, the authors postulate that hybrid methodologies are ideal for analysing textual data related to climate change discourse. The authors also demonstrate that text mining is an open and evolving field, in the sense that it can be combined with other approaches to shed new light on the climate discourse.

DOI: 10.4018/978-1-6684-8634-4.ch004

1. INTRODUCTION

Climate change is an extremely complex global phenomenon but is often experienced locally. It has gone from being analysed as a mainly physical phenomenon to being a social, cultural, political, ethical and communicational phenomenon (Zaccai 2012, Moser 2016, Pearce, Grundmann et al. 2017, Atwoli, Muhia et al. 2022, Kumpu 2022, Toivonen 2022).

Understanding the scope of the problem and leading thinking of identifying and managing key elements of the problem is challenging given the wealth of academic and grey literature and other bodies of knowledge that exist on a given topic and that often grow exponentially. The sheer volume and diversity of data make it difficult for human experts to process it and gain meaningful insights.

Text documents play a significant role among existing information resources. Analysis and processing of such resources are relatively simple tasks for an expert, while automation is difficult. The main reasons for this are the low coherency of text structures and a lack of obvious methods for their interpretation.

Consequently, communication related to climate-related risks and policies is becoming a major challenge for experts in various fields, including economics, discourse analysis and, in particular, language and computer sciences.

In the many debates that take place in different areas, one observes a multitude of voices, points of view, values, and interests related to climate change. Many actors seek to identify risks, formulate key questions and decide on priorities for action. Their points of view may vary from country to country and from company to company.

Language plays an essential role in the conceptualization and framing of discourses related to global climate change risks. These discourses continue in various text documents and constitute a heterogeneous material, occurring through different types of disclosures.

In this chapter, given this heterogeneity in climate disclosures, we postulate that hybrid methodologies are ideal for the analysis of textual data related to the climate change discourse. We would also like to show that text mining is an open field, in the sense that it blends with other approaches like machine learning and, thereby, helps to shed new light on the climate discourse.

Text mining is defined here as a set of concepts, methods and algorithms for processing textual resources that facilitate the automated processing of documents written in natural languages. It covers the field of data mining methods and techniques and combines them with the logic of text content analysis (Jo 2019). This approach is interdisciplinary as it draws on knowledge from fields such as:

a) Data mining (DM),
b) Machine learning (ML),

c) Natural language processing (NLP),
d) Information extraction (IE) and information retrieval (IR),
e) Statistics,
f) The theory of algorithms and data structures.

The relationships between these areas of research are presented in Fig. 1.

At this point, it is worth briefly discussing the similarities and differences between text mining and each of the fields as specified in Fig. 1.

One can easily notice the connections between text mining and *data mining*. In both cases, an analyst tries to find previously unknown regularities by exploring big data resources. However, the characteristics of processed resources are different: the data mining approach is usually adapted to the analysis of data with a fixed structure and with numerical values expressed on standard measurement scales. Text mining analysis, however, is oriented toward the analysis of texts; i.e., a resource without a fixed structure or with a structure defined in a vague and somewhat arbitrary manner. Text mining draws on data mining extensively—it adopts the exploratory research method of the analysis process, and the applications of the analysis process.

Figure 1. An overview of the disciplines that contribute to the field of text mining

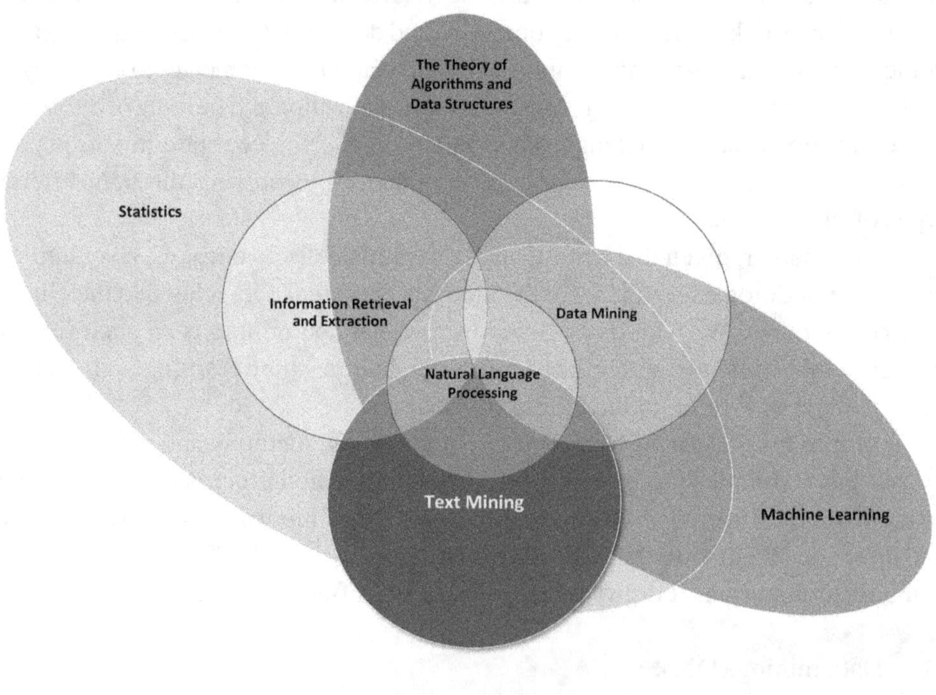

As with data mining, the use of the above-mentioned mechanisms makes it possible to discover previously unknown conditions, rules, sequences or patterns in text documents. Thus, discovered knowledge representation consists of creating descriptions of the empirical world or its states by utilizing data processing and analysis methods and techniques, particularly inference procedures.

Text mining shares many characteristics with *machine learning*, which (like text mining) would be impossible without utilizing methods and tools of advanced statistics (Rybak, Hassall et al. 2022). Machine learning mainly aims to detect existing regularities by analysing multiple examples and then encapsulating them into a model that can be easily processed by a computer system. The most commonly used model types include neural networks, decision trees and genetic algorithms. Machine learning is increasingly used in text mining. Further discussed advances in text mining as well as case study were made possible by supplementing text mining methods with machine learning techniques.

Text mining also benefits greatly from recent developments in *natural language processing* (NLP). The pioneer in this field is considered to be Noam Chomsky, an eminent linguist. He was convinced of the existence of general rules allowing the creation of such a mathematical (formal) model that would make it possible to comprehend and create correct sentences in natural language (Sperlich 2006). Chomsky reasons that grammar is a set of symbols and corresponding rules that determine their processing. Despite the great achievements of formal languages (evident, for example, in the compilation techniques used in computer programming), the current knowledge indicates that natural languages still elude the boundaries imposed by their formal description. Chomsky's successors continued his work, covering an increasing range of topics (McGilvray 2014).

The study of natural language processing generally started in the 1950s, although some work can be found from earlier periods. The early NLP researchers, among other issues, were interested in the use of natural language as a means of communication between humans and machines. Fulfilling this postulate requires a computer to have the ability to analyse human language and generate utterances on its own using the same mode of communication. The computers' ability to use natural language was of such great importance that Alan Turing considered it a sufficient criterion for attributing the characteristics of an intelligent being to a machine (Turing and Haugeland 1950).

While debating the issue of using a machine to analyse using natural language, one should also mention the achievements of Joseph Weizenbaum, the creator of the ELIZA programme, which can hold a conversation (via keyboard and monitor) on seemingly any subject (Weizenbaum 1966). The ELIZA programme became the prototype for many further projects aiming to create even better language processing tools. However, like all subsequent similar projects, the ELIZA programme has not achieved a human-level natural language capability. Thus, no project has met the basic Turing criterion–a necessary condition for attributing the characteristics of an intelligent being to a machine. One can find some significant similarities between the implementations of text mining analysis and NLP. In both cases, a computer system must be able to understand textual information. A typical NLP system is expected to respond properly to every utterance, while a text mining system is focused on analysing and processing textual resources in large amounts. In the latter case, the system is not expected to generate results in a natural language form (often, on the contrary, fixed format data are desirable to appear in the output).

Information Retrieval (IR) is the process of searching for and locating textual information in a database based on a user query. Searches can be based on a document's existing metadata, its full text or its content indexing. There are two main methods of document indexing and retrieval that are most commonly used, which are based on Boolean algorithms and rankings (Roshdi and Roohparvar 2015, Buttcher, Clarke et al. 2016). According to the Boolean model, an algorithm (based on words connected by logical operators) divides the text dataset into two parts: *matched* and *mismatched* to the search query. The ranking system—similarly to NLP—usually uses vector space models (Abualigah and Hanandeh 2015, Ogheneovo and Japheth 2016) and artificial neural networks (Mitra and Craswell 2017, Guo, Fan et al. 2020) or other probabilistic models (Zhai and Lafferty 2017) to assess the similarity between the content of the query and the content of text documents. It then determines the order in which the documents match the search query. Information retrieval has the advantage of being independent of any particular knowledge system or user query language.

Text mining is quite often equated with advanced methods of IR (used, for example, in online search services). However, such a perspective is not accurate, as the two mentioned forms of analysis have different purposes. The purpose of these search methods and tools is to return a set of documents that meet given criteria, while text mining is designed to provide information extracted by document analysis, not the documents themselves. In this context, the way in which the information representation mode is obtained by analysis should be considered to be an essential issue.

Information extraction (IE) is the process of identifying and extracting content from documents written in natural language and is based on analytically generated or predefined knowledge patterns (Moens 2006). These patterns are based on NLP-like solutions and use domain knowledge generators (content patterns) operating on the lexical analysis of textual data (Mannai, Karâa et al. 2018). Information extraction systems do not search for documents but for their content.

Information extraction research has been ongoing since the late 1980s and early 1990s. Between 1987 and 1998, multiple studies have been funded by DARPA (Grishman 2019). The aim was to develop methods of extracting well-defined information and then embedding it into a pre-designed framework structure (a template). The approach was used to analyse transmitted military messages, agency and economic news, and to detect signs of potential illegal activity. The proposed solution's focal points were mechanisms allowing the identification of proper names and coreferences (i.e., different expressions referring to the same object). These methods are currently among the most popular text mining algorithms and their field of application is constantly expanding.

Text mining, like any field of data analysis, draws on centuries-old achievements in statistics, which–in relation to the observed reality–should describe, foresee and explain regularities. Against this background, text mining can be regarded as an area of applied statistics that is unique and challenging in many places. Descriptive statistics methods, Markov chains and probabilistic models are especially popular among researchers of linguistic phenomena.

In the area of text mining, *statistics* denotes the application of quantitative methodologies to extract, scrutinize, interpret, and organize data procured from textual sources. This encompasses a spectrum of techniques, ranging from rudimentary frequency analysis, where the occurrences of specific words or phrases are enumerated, to more intricate procedures such as co-occurrence and collocation analysis, which evaluate the relational dynamics and proximities between textual elements.

Advanced statistical methodologies in text mining may also involve probabilistic models like Latent Dirichlet Allocation (LDA) for topic modeling and classification tasks. These sophisticated techniques facilitate the extraction of latent patterns, trends, and structures within voluminous text corpora, thereby enabling the transformation of unstructured text data into actionable insights.

The ultimate objective of employing statistical techniques in text mining is to distill meaningful and interpretable information from large-scale text datasets, which can subsequently be utilized to inform strategic decision-making processes, enhance understanding of relationships, identify emergent trends, and contribute to the broader field of knowledge discovery in databases (KDD).

When discussing the text mining domain, one should also mention the field of the *theory of algorithms and data structures* (algorithmics). Using computers to process textual data has a relatively long history beginning in the 1950s. Novel algorithms for text data processing have been used by the greatest authorities in algorithmics (Weiss, Indurkhya et al. 2015). Modern text mining makes extensive use of computer science advances in the design, analysis and processing of data structures. These tools are capable of solving problems related to the way in which computer systems store acquired information. Many text mining problems are well known to data structure experts. At the boundary between these fields, the issues of storing large amounts of text, implementing indexing operations, or defining queries have been researched.

Notwithstanding the considerable research corpus that exists pertaining to text mining applications in various areas, it is evident that the domain of climate change remains largely under-represented, as pointed out in *Tackling Climate Change with Machine Learning* (Rolnick, Donti et al. 2022). The authors contend that the application of machine learning and natural language processing strategies holds substantial potential to enhance this field of research. A proposition was advanced for conducting research on climate disclosures employing cutting-edge text mining and NLP tools (Luccioni and Palacios 2019). However, as of the current writing, this proposition has yet to be appropriately realized. The mounting significance of such disclosures, coupled with the intrinsic challenge of their collection and analysis owing to their diverse and scattered nature, suggests that the process of investigating these disclosures *en masse* could greatly benefit from elements of text mining methods. Therefore, the next sections of the chapter will describe:

- The detailed definition and objectives of text mining applications and their links to machine learning and other related fields,
- Different types of problems considered within text mining, with a brief discussion of exemplary applications,
- Methods of representing the information contained in text documents,
- Basic methods for processing information extracted from text documents,
- Practical analysis of a sample set of documents to illustrate the discussed issues

The conclusion of this chapter is a methodological reflection on the process of developing computer-aided analysis of text data, which progresses towards machine learning-enhanced text mining methods for knowledge discovery.

2. METHODS

Text mining involves several phases (see Fig 2):

1. Text data gathering
2. Text data pre-processing
3. Text data representation
4. Definition of the purpose, scope and costs of the analysis
5. Content analysis/implementing algorithms
6. Interpretation and integration of analysis results

Phases 2-5 are described in detail next.

Figure 2. Stages of the text mining process

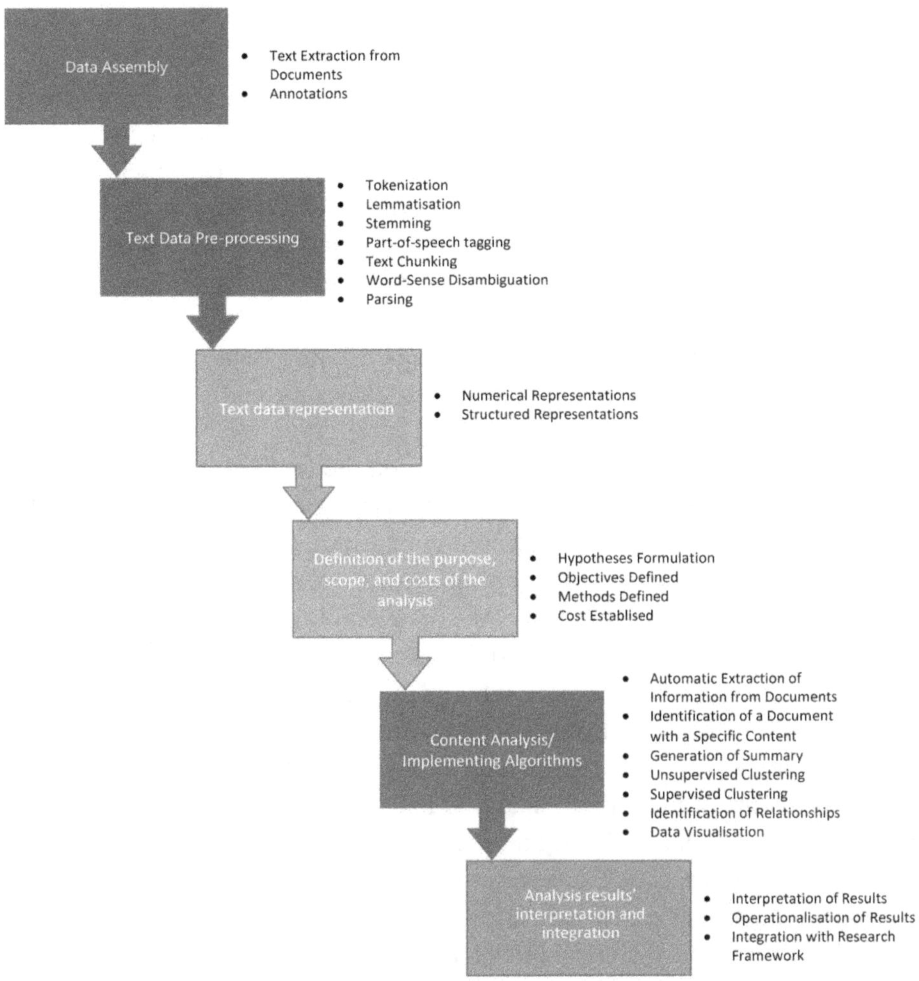

2.1 Text Data Pre-Processing Methods

The first phase in the implementation of text mining analysis is the pre-processing of a set of documents, during which textual data saved in different formats are imported into a single batch that is simple to read and convert. Any "raw" text document subjected to data analysis has to be converted into a specific format.

That requires the conversion of the textual data structure into one more suitable for the subsequent stages of the text mining-based analysis process. The text mining methodology assumes that each text document is represented by a limited set of words organised by syntactic structure. The processing of a document prepared in a natural language involves syntactic reduction of the text document, which excludes components irrelevant to the analysis process.

The pre-processing of documents includes the following steps:

a. Conversion of documents to a text form (a large number of programs and libraries capable of text analysis can only operate on specific text files),
b. Removal of document formatting (e.g., formatting tags),
c. Unification of character coding (this step is crucial for texts containing diacritics, as the coding of characters specific to different languages is not unified).

As a result of these pre-processing steps a batch of text files is obtained that contains textual data with standardised character coding and without any formatting information.

The subsequent analysis process depends on the method used for textual data representation. When choosing a representation method based on a list of words in a document, one can consider the most common techniques currently used to prepare textual data: tokenization, stop words, lemmatisation, and stemming.

Tokenization of a text document is the division of the input text into sentences, words, punctuation marks (commas, full stops, etc.) and non-text characters. Other elements of a document that have the status of regular expressions can also be used as tokens. Tokenization is usually an automatic process, depending only on the formal language of the text mining programme. The results of tokenization are: creating a set of words that occurs in all analysed text documents, removing punctuation marks from subsequent analysis, and replacing other non-text separators with so-called "single white spaces". The series of expressions obtained by combining all the text documents forms a classification dictionary that is further processed during the development of the analytical model. At the same time, to reduce the dictionary volume (and, thus, the dimensionality of the description of the batch of analysed text documents), a further reduction in the set of words and phrases describing the documents is made. At this point, it is worth noting that there are several other methods of text dimensionality reduction (Kannan, Gurusamy et al. 2014, Vijayarani, Ilamathi et al. 2015); however, due to the scope of this chapter, only the ones most commonly used in the text mining approach are described. It should be stressed that the usefulness of other methods in text mining depends on the aim of the analysis or research, and should be evaluated individually each time.

Natural languages contain multiple auxiliary words which, from the perspective of text mining analysis, do not convey any relevant information about the content of the text document. These are usually conjunctions or functional expressions. The next step is to eliminate them by creating so-called *stop words* for further syntactic reduction of the textual data. Such a list of words is not limited, and a researcher may add other words considered irrelevant to the analysis of the textual data. Stop words are closely related to the language of text documents. Skipping functional words is a simple way to reduce information noise and improve the quality of text representation, and has long been used in almost all applications related to information retrieval and text mining. Stop words can be established (a) manually, based on knowledge and experience, (b) automatically, based on the frequency of character strings found in the text database, or (c) automatically with supervision, when a researcher verifies the automatic operation.

The next phase of text processing is *lemmatisation*, i.e., morphological analysis of the dictionary and reduction of similar lexical forms of words to one basic form – a *lemma*. A lemma is the canonical, simplest form of a lexeme used for its dictionary representation. A lemma can be represented by a single text word. In spoken languages, the inflectional variety of words is essential in constructing a syntactically correct sentence. It also indicates the function of a word in a sentence. However, the correct representation of a text document requires that lexically similar words should be recognised as occurrences of the same lexeme as they convey the same information. The lemmatisation process is particularly important for languages with a variety of inflections.

In the process of mining text documents, it is not necessary to find the correct lexical form of a word to calculate the similarity between two documents, as it is enough to find its stem; i.e., a part of it that enables the identification of a lexeme. The process of reducing a word to its core is referred to as *stemming*. Stemming is the reduction of a group of words to their main stem, a basic form that makes it possible to count them as the same word. A classic example of this procedure is Porter's algorithm (Porter 1980). Algorithms like this are usually based on rules constructed specifically for a particular language. In practice, reducing a word to its stem means removing its characteristic prefixes, suffixes and declensions while preserving its meaning. The stemming process results in a classification stem development for the words in the dictionary.

When choosing a method of representing data in a structured form, it is essential to consider the semantic context. Hence, depending on the required research goal, one can also apply a method that assigns words to the specific parts of speech in addition to the previous steps of preparing textual data for analysis. That is because nouns, verbs and adjectives have different meanings in a text. For example, nouns describe specific objects, events and facts, and also indicate the subject matter. Adjectives, on the other hand, indicate the qualities and characteristics of nouns. Similarly, the textual data can be reduced based on a document's sequences of words referring to single entities, such as persons, institutions or organisations. This is because word sequences facilitate the correct representation of words, especially ones with a specific meaning. Neglecting this type of contexts in the analysis may lead to semantic misinterpretation of textual documents. To further reduce the number of words in a dictionary, one can use indexing or keyword search algorithms.

A simple way to verify keywords is to use their *entropy*, which is an *information theory* measure that expresses the average amount of information conveyed by a single message. It can be described as the uncertainty involved in the occurrence of a given elementary event at the next instant. If an event occurrence probability equals 1, its entropy equals 0 because it is known in advance what will happen—there is no uncertainty. In text mining analyses, the entropy emphasis is on rare words that only occur in a few documents in an entire batch. Such words are assigned the greatest weight of all words in the batch. If they occur in multiple documents, the entropy rate will be low. Entropy can be perceived as an indicator of the meaning (weight) of a word in a given domain or in the context of textual data analysis.

Besides the methods for preparing textual data for text mining described above, one can also use NLP pre-processing measures to enhance the representation of the semantic context in text documents. These measures include:

1. Part-of-speech tagging (POS) – tagging parts of speech according to the type of language involved,
2. Text chunking – dividing a text into smaller analytical units in order to group adjacent words or words in a sentence so that it is easier to find phrases,
3. Word-sense disambiguation – semantic categorisation to determine the meaning of single words or phrases in documents for a better vector space representation, and
4. Parsing – syntactic analysis of a text document, its grammatical structure and its correspondence to natural language grammar.

3. TEXT DATA REPRESENTATION

3.1 Numerical Representations: Machine Learning and Vector Space Model

In the text mining analyses, similarly to IR and NLP, currently the most popular method of numerically representing text documents is a machine learning-based method—the vector space model (Salton, Wong et al. 1975, Erk and Padó 2008, Schütze, Manning et al. 2008).

In this model, each document is represented by a vector embedded in a so-called *feature space*. The vector representation of documents enables mathematical conversions, which are considered to reflect operations on actual documents. The vector space model also allows the retrieval of documents containing specific words with vectors close enough to the query vector (the query vector is assigned to the vector defining the document). This model is usually represented as a document matrix of words/set of words (the so-called *bag of words*), where the value is assigned based on the number of occurrences of a word in a document. The order of occurrence of words in the text is not measured. The matrix can also reflect the relations between documents and word strings with an *n* number of words (*n*-gram representation) or between documents and concepts, ideas or facts (conceptual representation), whereby it is necessary to identify the concepts in the documents.

The vector representation model assumes that a text document is represented as a *keyword frequency vector*, and a batch of documents can be represented by a matrix (*term frequency matrix*) in which the elements represent the number of occurrences of a given keyword in a given document.

Each element of the matrix represents the weight of a word in a document. In the simplest Boolean representation, in a document vector, the word weights have the values 0 (not present) or 1 (present). In its full representation, these weights correspond to the frequency of the words in the documents. Each element of the vector represents a word (or group of words) in the batch of documents, and the size of the vector is determined by the number of words (or group of words) of the complete batch of documents.

The advantage of the vector space model over a representation based on a set of keywords is that it allows the definition of a measure of distance between the documents and a query. Documents with similar content should have a similar frequency of the same keywords. The vector representation of documents allows each document to be interpreted as a point in a high-dimensional space whose dimensions correspond to keywords. Hence, Euclidean space distance measures can be used to assess the distance between documents or between the documents and a query. Information retrieval systems also use specific measures such as cosine similarity or word distance measures (Schütze, Manning et al. 2008).

As mentioned earlier, a significant feature of vector representation is its ability to determine the similarity of documents with respect to each other (Schütze, Manning et al. 2008). This method of constructing a document matrix is based on a simple weighting of documents according to the frequency of keywords in the "preferred" documents that are more likely to have a particular word in the search query. Their ability to differentiate (discriminate) text documents is low. In practice, text documents are more efficient in describing words whose occurrence frequency is low. Therefore, the TFM model adopts a document weighting scheme that considers the discriminatory features of keywords.

This scheme is called TF-IDF, where TF stands for the *term frequency* weight and IDF means the *inverse document frequency* weight. The TF weight is the number of occurrences of a word in the document, while the IDF weight is the logarithm (decimal or natural) of the quotient of the total number of documents to the number of documents containing the keyword. The TF-IDF value of a keyword in a document is the product of the two weights. The TF-IDF measure has a higher discriminatory capability for text documents than the classical weighting system of simple TFM (Schütze, Manning et al. 2008).

From the perspective of text data mining and knowledge discovery, the representation models should pursue the maximal perseverance and reproduction of the semantic content of the document and provide an efficient mode of information retrieval. It should return its measurement of similarity to the content defined in the user query. In the text document representation models based on a set of keywords, this problem relates to the issues of *synonymy* (expression of the same content using interchangeable words), *polysemy* (the occurrence of different meanings of a given word in different contexts), the way keywords are defined (singular or plural), or the declension of words in certain languages. For vector representation models, this issue is related to the capability of grouping text documents based on measures of discrimination and similarity.

Practically, a process of retrieving and mining text data using a vector space model can be divided into three stages: indexing keywords in documents, weighting the indexed words for retrieval, and ranking documents according to the assumed measure of similarity.

3.2 Structured Representation: Conventional and Machine Learning Approaches

The main advantage of the method of constructing machine learning-based numerical representation of texts (presented above) is its simplicity in implementing the calculations and wide range of methods necessary for further processing of the obtained structure. Its disadvantage is its inability to consider much of the information available in the source text. The reason for the loss is considering information only about the occurrence of the words, without analysing the linguistic structures they create.

The basic idea is to counteract the loss of information by using data structures suitable for storing relations resulting from the word order, the characteristics of objects described in documents and their relations, and the cause-effect relations. Structures convenient for implementing these postulates can be strings, lists, records, trees and graphs.

Unfortunately, the application of such a solution may encounter a number of problems. The first is the difficulty in converting a text document into a structured form. The second is the much smaller range of analytical methods that can be adapted to deal with stored information that uses complex structures.

Early efforts in structuring the information extracted from documents aimed to place it in table records embedded in relational databases. This implementation enables further exploration of information by means of well-known and very efficient query mechanisms based primarily on the SQL language.

Another approach advocates moving from relational to tree-based structures and storing information as XML-defined documents. This concept takes advantage of XML's main feature, which is its ability to define its own tags and precisely specify the relations between them.

The use of this approach requires the assumption of an identical domain scope of all simultaneously processed texts and the existence of formal tools describing the reality for the domain of interest. Currently, the focus of this research is on the subject of *ontology*, which is a formal way of modelling a specific fragment of reality. An example of pioneering work in this field is (Gruber 1993).

The definition of ontology includes the description of objects that exist in reality and their relations. The objects are represented by data structures called *entities*. They usually have a hierarchical structure that is represented in the form of tree structures. Among the relations between objects represented in an ontology, semantic relations (types: has, is a component of, etc.) play a particularly important role. For this reason, the arrangement of objects together with a description of their semantic relations is called a *semantic web*. Graphs are convenient structures for representing semantic webs.

Unfortunately, converting the information contained in a text into a structured form (even as a simple table) is generally challenging. This process can be carried out manually but is very time-consuming. It can be done using rules defined by experts but, usually, the rules only cover the most typical cases and are only able to correctly process small amount of information when operating in a layout even slightly deviating from the standards established by an expert. **Currently, various machine learning techniques are the most promising way to implement this process, however it requires comparatively large input sets.**

The use of a combination of both methods—structured representation of text data combined with representation based on a list of words occurring in a document—is particularly promising (yet relatively complex).

In this case, a representation method based on a list of words occurring in a document can be used as a tool to support the construction of ontologies by using statistical analysis of texts to identify the most relevant objects and relations (while conducting this type of task, it is necessary to use external dictionary information).

An example of the effective combination of these two machine learning methods—numerical and structured representations—will be presented in *Section 7. Case study: Machine Learning-enhanced Text Mining Analysis of Climate Change Disclosures.*

4. DEFINITION OF THE PURPOSE, SCOPE, AND COSTS OF THE ANALYSIS

Defining the purpose of a text mining application is one of the most important steps in the analysis process. The completion of this stage determines the entire course of further work. The purpose affects the form of hypotheses formulated, the set of methods used, and the way the results are presented. The scope is a product of the purpose, the methods chosen and the time available. Costs can be of a diverse nature. Their careful analysis provides a basis for deciding whether to carry out text mining-based analysis, or whether to extend or limit its scope. It is particularly important to compare the costs of the analysis with the potential benefits of its results.

5. CONTENT ANALYSIS

The range of text mining applications for content analysis is particularly wide. The attempts to identify the most important fields of application can be found in multiple sources (e.g., (Fan, Wallace et al. 2006, Gupta and Lehal 2009, Aggarwal and Zhai 2012). The discussion presented in this part of the chapter is certainly not complete and is only an overview of the most widespread, machine learning-related applications.

5.1 Automatic Extraction of Information From Documents

Information retrieval mechanisms are based primarily on attempts to match patterns with particular parts of a text that determine the type of content being sought. The simplest type of pattern can be a set of keywords. Potentially, one can create patterns to identify phrases specifying a type of event, time, location, performer, effects, etc. The information extracted in this way, after being given a strictly defined structure, can be placed in traditional databases for future, convenient processing.

Recently, a dynamic development of this type of application can be observed in relation to the analysis of reports in the field of medical research (Korhonen, Ó Séaghdha et al. 2012, Sun, Cai et al. 2018). This field has seen an exponential increase in publications over the last few decades. Utilising machine learning-enriched text mining tools makes it impossible for researchers to familiarise themselves with the most up-to-date literature in their specific area of interest.

Therefore, ways to automate this time-consuming process are greatly desirable. The primary objective of the work in progress is to create a system for enhancing the process of retrieving information from expert literature, giving it a defined structure and placing it in a database (Korhonen, Ó Séaghdha et al. 2012) that facilitates further processing.

Similar work is yet to be conducted in the field of climate change literature.

5.2 Identification of a Document With a Specific Content

In this case, the aim of the analysis is to create a system that monitors a large number of documents and identifies those that may be relevant to a criterion. It is assumed that the identification of potential documents is automated but not their analysis, which should be done by a human expert.

The work presented in (Bouaziz, Nabli et al. 2019) examines potential applications of this type of system to enterprise information systems, where it is used as a basic tool to process information in textual data storage.

This type of storage should collect, among other things, reports, created documents, notes, correspondence, and press releases. Text mining methods enable searching for information, retrieval of documents according to defined criteria, determination of the documents' nature (subject matter, key topics, language), organisation of the documents' structure, visualisation of collected resources, and construction of a customer profile (by analysing queries, requests and comments sent by a customer).

5.3 Generation of a Summary

There have been exponential increases in document numbers in almost every field, which makes it virtually impossible to familiarise oneself with them all. A partial solution to this problem is the automatic generation of document abstracts (Verma and Nidhi 2017).

Attempts to automate an abstract generation system are among the main text mining applications. This is a complicated task since it requires extracting the most relevant facts from a document (often with a previously unknown structure) and expressing them in properly formulated sentences in natural language.

The simplest solutions identify the most relevant keywords using statistical methods, but such solutions do not provide information about the nature of the relations between particular words. Another approach provides a summary of key sentences taken directly from a source text. In this case, there is a need to select sentences that best represent the document under analysis. Various attempts have been made to determine the weights of sentences. The first group of solutions aggregates the weight indices for individual words that create a sentence and takes the resulting value as an assessment of the sentence's relevance. The sentences are then ordered by decreasing information value. The most relevant ones are included in the abstract. Another approach includes only those sentences that contain certain phrases indicative of high informational value in the summary. Examples include: "greatest achievement", "summarising", "significant results", etc.

5.4 Supervised Classification

The purpose of model-based classification is to analyse a set of documents and assign each of them to one of the previously identified classes while considering the information they contain. This requires defining the models of each class and measuring the similarity between each model and the analysed document to determine the best match (Kadhim 2019). The key to how these tasks are carried out is the way in which the information contained in the document is represented.

5.5 Unsupervised Classification (Cluster Analysis, Grouping)

In the analysis of text documents, unsupervised text classification—also known as *model-free classification*—involves determining the homogeneity of a document batch, defining groups of similar documents, determining their relations, and characterising their features (Shafiabady, Lee et al. 2016). The most relevant problem in solving cluster analysis issues, as with model-based classification, is to determine a measure of similarity between documents. This is a prerequisite for the implementation of cluster analysis using the most common classification algorithms (e.g., taxonomic methods, artificial neural networks, etc.).

5.6 Identification of Relationships

Identification of relationships can be understood as 1) the detection of the relationships between the information extracted from different text documents; or, 2) the identification of documents that are related by the content they contain.

In the first case, there are two main stages. The first stage consists of extracting information from the source document, and the second involves identifying links between the extracted information The latter stage simply requires identifying a set of documents related to the document or subject concerned. It is assumed that further analysis will be performed by a human.

Various link identification systems have been presented in the literature. A good example is the system for automatically analysing claims reports to detect the causes of increases in claims reported by customers (Sabban, Lopez et al. 2020). The information extracted from the texts, used in conjunction with quantitative data, can identify weaknesses in existing solutions and detect potential sources of fraud.

5.7 Visualisation

Visualisation is usually used in conjunction with other type of text mining tasks. Its main purpose is to provide the user with a simple method of interpreting the results obtained. Most often, visualisation is done to present relations between extracted entities and events or relations within the structure of a batch of documents (Don, Zheleva et al. 2007). Currently, knowledge graphs play a crucial role in presenting information obtained by text mining and machine learning processes (Pujara and Singh 2018).

5.8 Automatic Answer Generation

This type of issue, which is of broad interest in the field of artificial intelligence, concerns research into the ability of a machine to understand a question posed by a human in natural language. Ongoing work in this field aims to determine the feasibility of using natural language as a tool for embedding an interface into advanced information systems (Clark, Pantidi et al. 2019).

6. IMPLEMENTING ALGORITHMS

When proceeding to this fundamental stage of machine learning-enhanced text mining analysis, it is important to choose the most appropriate solution. The primary consideration should be the purpose of the analysis. The way in which information from textual resources is represented should also be reviewed.

6.1 Singular Value Decomposition Method

One of the difficulties in the analysis of a frequency matrix (in its original or converted form) is its large size as its number of rows is equal to the number of different words in the examined documents. The *latent semantic analysis* (LSA) method described in (Dumais 2004) is a widely applied solution. It applies a distribution according to singular values to the frequency matrix (in its original or converted form).

The purpose of the calculation is to define a space in which it is possible to analyse sets of words occurring in documents (words are represented by a U matrix) and to analyse a set of documents (a V matrix). Each object (a word, a document) is represented by one row of the appropriate matrix, where (as in the analysis of principal components) the individual coordinates are ordered in descending order according to their information value. This reduces the space required by limiting the matrix to a certain number of initial elements of the vector.

When dealing with document classification, the presented scheme is applied to the V matrix. Its elements, when scaled by the coefficients contained in the S matrix, determine the points representing the documents of the study. From a practical perspective, it would be particularly useful to reduce the space in which the documents are described to two or three dimensions, as this would allow the structure of the document batch to be represented graphically.

The LSA method is a convenient tool for processing information represented as a word list. Its results are usually the starting point for further computations.

6.2 Taxonomic Grouping Methods

Taxonomic methods are among the traditional methods of data analysis. They are used to group objects defined by an adopted set of attributes. A distance matrix or a similarity matrix derived from the analysed objects is usually used as input data.

Taxonomic methods can be applied to the analysis of information retrieved from text documents, as long as a way of determining similarity between documents is available for the adopted representation and can be expressed by a numerical value. Among the taxonomic methods, hierarchical methods are a special group that enables the detection of an existing hierarchy in a batch of documents. Results obtained this way can be used as the foundation for a graphical representation of the structure of the analysed collection. (Yang and Liu 1999) provide an interesting comparison of taxonomic grouping methods.

6.3 Classification Trees

Classification trees can identify rules to classify objects into predefined classes. The results of the algorithm are presented as a decision-making tree or set of rules. Classification trees are often used to find regularities in databases and data storages. In text mining analyses, these types of algorithms are particularly useful if the information retrieved from documents has been structured beforehand.

6.4 Neural Networks

Neural networks can also be successfully used to analyse very large sets of document data. The richness of neural models allows them to be applied to supervised and unsupervised classification tasks and to describe highly intricate non-linear relations that exist between the acquired information. A review of the application of a specific subgroup of artificial neural networks – deep neural networks (DNNs) to text mining can be found in (Minaee, Kalchbrenner et al. 2021), while the application of modified DNNs to the identification of hierarchical relations is presented in (Kowsari, Brown et al. 2017). Noteworthy application of DNNs in text analysis is *anomaly detection* (Rybak and Hassall 2021).

6.5 Support Vector Machines

The *support vector machine* (SVM) method constructs a hyperplane separating the examined objects. The position of objects is not taken into account when determining the boundary between them but is only limited to those in the closest vicinity (the coordinates of these points define the so-called *support vectors*). If it is not possible to make divisions by means of a hyperplane, objects are immersed into a more dimensional space in such a way that the new points feature linear separability.

Applying the SVM method to text mining is of particular interest due to its adaptability to different applications. SVM has been successfully used for the supervised classification of documents over the last two decades when using representation via word frequency information (Basu, Walters et al. 2003) and complex structures (Gartner 2008).

6.6 Association Analysis

Association analysis can also be useful in processing information extracted from documents. It is a convenient tool for identifying regularities in structured data. The results of the analysis can serve as a basis for visualising existing relations (Wong, Whitney et al. 1999).

7. CASE STUDY: MACHINE LEARNING-ENHANCED TEXT MINING ANALYSIS OF CLIMATE CHANGE DISCLOSURES

To illustrate how text mining analysis supported by machine learning techniques results are interpreted and integrated, an end-to-end pipeline for the automatic extraction of information from documents via deep neural networks and data visualisation is presented. This analysis exploits the ability of machine learning tools to assess large textual data resources such as CRDs and thereby extract underlying relationships and themes. The specific task is to validate whether text mining can automatically extract risks, assign their types, and provide insightful data visualizations. Risk in this context refers to the uncertainty that can have both positive and negative impacts on achieving objectives. Companies now face the direct risks of climate change as well as the transitional risks associated with significant changes in human productivity. The push to address climate change risks is being driven by investment and trade, regardless of supportive policies. As a result, companies are adapting their risk management practices to effectively respond to these new pressures.

To be seen as responsible actors in the climate crisis, companies are conducting climate risk assessments and making disclosures. Climate risk disclosure (CRD) involves reporting on the identification of material climate risks and outlining efforts to manage them (Sanderson, Irato et al. 2019). The form of disclosure varies and is mostly voluntary in most countries. To ensure the rigor and usefulness of these disclosures, the Financial Standards Boards developed the Taskforce for Climate-related Financial Disclosures (TCFD) reporting framework, which has become the industry standard.

Australian companies face the same pressure as international firms to manage climate risks. Australian regulators have started providing guidance on aligning financial disclosures with CRD as best practice. The Australian Prudential Regulation Authority emphasizes the importance of disclosing decision-useful, forward-looking climate risk information to assess an institution's resilience (Caldwell 2020, APR 2022). The demand for climate disclosure has increased as investors want to know if their investments are being managed with climate risks in mind. Therefore, it is crucial for firms to understand their exposure to climate risks and adopt best practices for managing them (Chipalkatti, Le et al. 2021).

Nonetheless, there is a very limited research on which climate risks and management responses Australian companies consider significant. The analysis of CRDs can provide valuable insights into climate risks and associated risk management. The analysis utilises the steps shown in Fig. 3.

Some 839 companies listed on the Australian Stock Exchange were surveyed for climate risk disclosures, with 201 such disclosures identified. Diverse reporting approaches have contributed to the variability in the extent and quality of Australian climate risk reporting. Accessing climate-risk disclosures in Australia often requires

Figure 3. End-to-end pipeline for automatic extraction of information from documents via neural networks and data visualisation

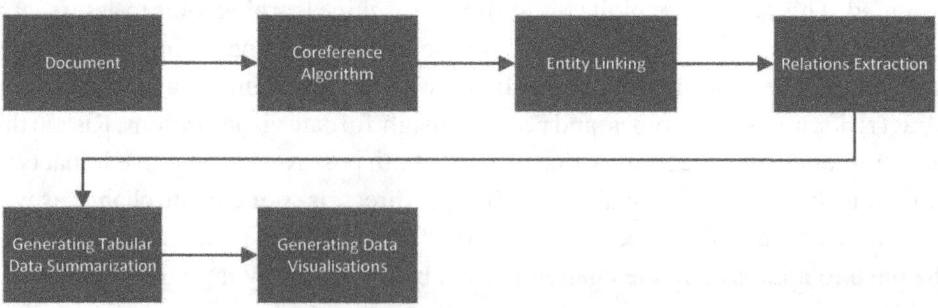

an extensive process that discourages casual scrutiny. Presented dataset was pre-processed as described in detail in *Section 2.1 Text Data Pre-processing Methods.*

Subsequently, to automatically extract information from the documents, a part-of-speech tagging (POS) and *coreference resolution* method from the SpaCy Universe library was used. SpaCy is an open-source software library for advanced text mining, machine learning and natural language processing that is written in the Python and Cython programming languages (Honnibal and Montani 2017). The coreference resolution method is an extension of POS and is used for finding all the expressions that refer to a specific entity.

Diving deeper into the specific techniques, part-of-speech (POS) tagging serves as the primary layer of analysis. It involves labelling each word in a text as corresponding to a particular part of speech, such as a noun, verb, adjective, etc. It's a fundamental aspect of modern text mining and is critical to many downstream tasks, such as named entity recognition, parsing, and automatic data visualisation.

Building upon POS tagging, the coreference resolution method adds another layer of sophistication. Coreference resolution is an advanced linguistic operation that identifies when two or more expressions in a text refer to the same entity. Essentially, it connects the dots between pronouns and the entities they reference in the text. For instance, in the sentence, "John said he would be late," coreference resolution identifies that "he" refers to "John."

Following the POS tagging and coreference resolution, text chunking, also known as shallow parsing, was performed. This process involves segmenting the text into smaller pieces, often into individual sentences or phrases, that still preserve the overall meaning. It's a step that helps in breaking down large pieces of text into manageable units, making it easier to handle and analyse. It also aids in extracting short phrases and relationships among entities within those phrases, providing rich, context-sensitive insights into the subject matter of the text.

These methods collectively create a powerful and precise system for mining valuable information from the documents, enabling to look deeper into the material and extract the relevant data needed for further analysis.

In the *named entity linking* part of the pipeline, we extracted all the mentioned entities and connected them to a target knowledge base which, in this case, was Wikipedia. Named entity linking is beneficial since it also deals with word-sense disambiguation; that is, semantic categorisation to determine the meaning of the words in a document for better vector space representation.

In the subsequent stage of the pipeline, *named entity linking*, a process was implemented where all cited entities in the text were isolated and linked to a target knowledge base. For the purpose of this case study, Wikipedia was chosen as the target knowledge base owing to its expansive collection of structured, crowd-sourced information.

Named entity linking, as an operation, is essential to text mining as it serves a two-fold purpose. First, it helps identify and categorize named entities mentioned within the text, such as people, organizations, locations, risks, and other distinct nouns. These entities are then linked to their corresponding entries in the knowledge base, thus providing an immediate context to the isolated terms. This linking transforms these entities from mere words in a document to unique, identifiable elements with specific semantic implications.

Secondly, named entity linking offers a significant advantage when it comes to word-sense disambiguation, a linguistic phenomenon where a single word may have multiple meanings based on the context. An example of this can be found in words like 'crane,' which could refer to a bird, a type of construction equipment, or an action to stretch one's neck. By employing named entity linking, each instance of the word 'crane' can be assigned to the correct meaning based on the context, thus improving the semantic categorization within the document.

The benefit of such semantic categorization is the creation of a more accurate vector space representation of the text. Vector space models represent words or documents as vectors in a high-dimensional space, enabling numerical computations that capture semantic or syntactic relations between words or documents. By disambiguating word senses accurately, we can ensure that the vector space representation utilized truly captures the intended meaning of each word, leading to more accurate and insightful downstream analyses.

Upon the successful extraction of the named entities, the pipeline proceeds to the next significant phase: *relationship extraction*. This part of the process serves to discern and establish connections between the various entities based on the context provided within the text. By doing so, it enhances our understanding of the interaction between different elements, thus enriching the depth and quality of the information derived from the document.

The methodology adopted for relationship extraction relies on parsing the text and iteratively examining every possible permutation of pairs of entities. By iterating through each permutation, the process ensures that no potential relationship between entities is overlooked, thereby resulting in a comprehensive mapping of relationships within the text. This exhaustive method is instrumental in inferring relationships even in complex texts where links between entities might not be immediately apparent, thereby enhancing the robustness of our analysis.

To facilitate this task, we utilized the OpenNRE library, a versatile tool for relation extraction that is equipped with machine learning-based models for this purpose. The OpenNRE library was used in this step, which features machine learning-based open-source relationship extraction models that were trained on the Wiki80 or Tacred datasets (Han, Gao et al. 2019). The key strength of this tool lies in its pre-trained models that are trained on benchmark datasets such as Wiki80 or Tacred.

Wiki80 is a widely-used dataset for relation extraction, which includes 80 different types of relationships extracted from Wikipedia. On the other hand, Tacred is a large-scale relation extraction dataset that includes a wide variety of entity and relation types, sourced from newswire and web text. Leveraging these pre-trained models ensures that our relation extraction benefits from the robust training these models have received on extensive and diverse data, thus guaranteeing the reliability of the relationships extracted.

The process of relationship extraction, facilitated by the OpenNRE library, not only allows us to map the intricate web of relationships between different entities but also ensures that the mapping is accurate and reliable, contributing to the overall validity and comprehensiveness of our analysis.

The extracted information structures can be visualized in abstractive form as shown in Fig. 4.

The specific visualization outcomes derived from the processed data are illustrated in Fig. 5. Automatic visualization of such data is a critical step in data analysis as it allows us to understand and interpret complex data in an accessible and intuitive manner using Neo4j framework (Lal 2015). Visual representations aid in identifying patterns, trends, and insights that might be difficult to discern from raw, numerical data. It plays an essential role in presenting the insights extracted from the dataset in a format that can be easily understood and communicated. Neo4j is a highly flexible and effective tool that is specifically designed for generating visual representations of data in a graph format. The strength of neo4j lies in its capacity to portray complex datasets with multiple entities and relationships in a visually coherent and organized manner.

The graph modelling enables to transform traditional tabular data into an interconnected network of nodes (representing entities) and edges (representing relationships), thereby mapping the intricate web of connections between entities and extracted risks within the dataset. With this technique, each entity becomes a node in the graph, while the relationships become the links that connect these nodes. This transformation from tabular data to a graph model makes it straightforward to visualize and understand the multitude of relationships within the data, particularly when dealing with large or complex datasets.

The presented example serves as an initial step towards constructing an effective decision-making framework. This framework is particularly pertinent in contexts like climate policy assessment, where consensus on probability distributions is elusive, and stakeholders exhibit a wide range of risk tolerances. A decision-making framework is paramount in these scenarios as it provides a structured approach towards evaluating multiple, often complex options, thereby facilitating informed, rational decision-making.

Figure 4. General visualization of named entity and relationship extraction using SpaCy and OpenNRE

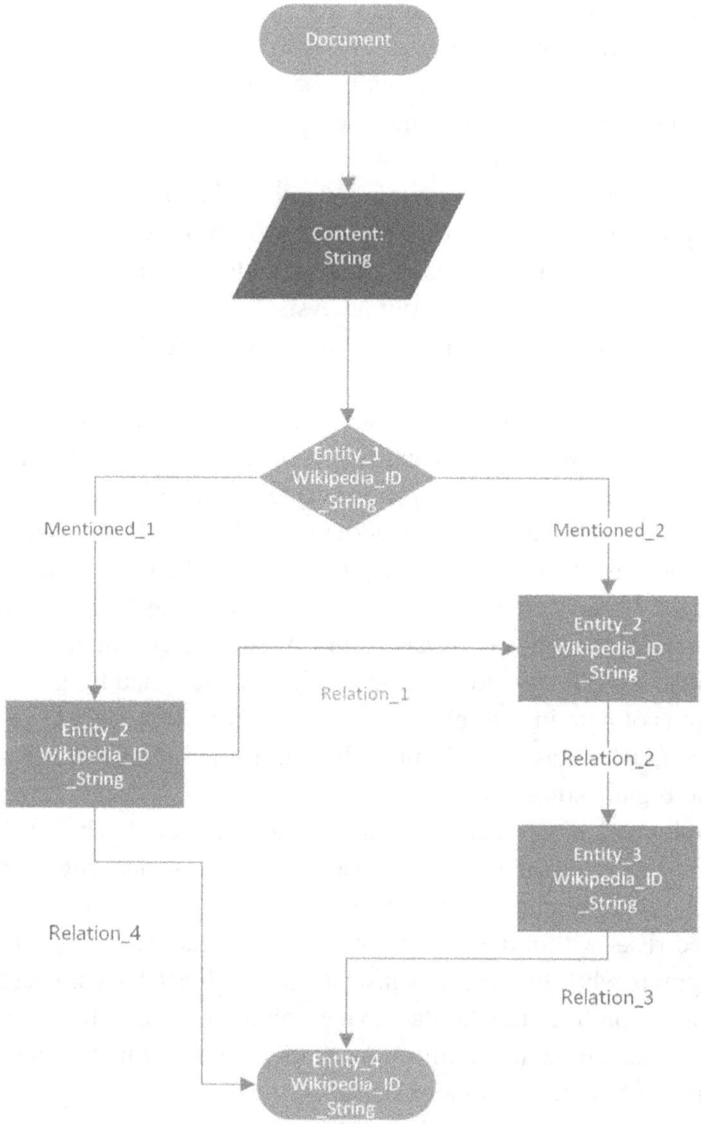

In the realm of climate policy assessment, decision-making can be intricate, largely due to the inherent uncertainties and the diverse interests of various stakeholders. Probability distributions, which offer a mathematical description of the outcome possibilities of a random event, can be challenging to agree upon, considering the unpredictable nature of climate change impacts. Further, the risk tolerance, or the

Figure 5. Data from 201 disclosures on climate risk visualized using Neo4j for graph data modelling. The risks and their types were extracted automatically using text mining methodologies

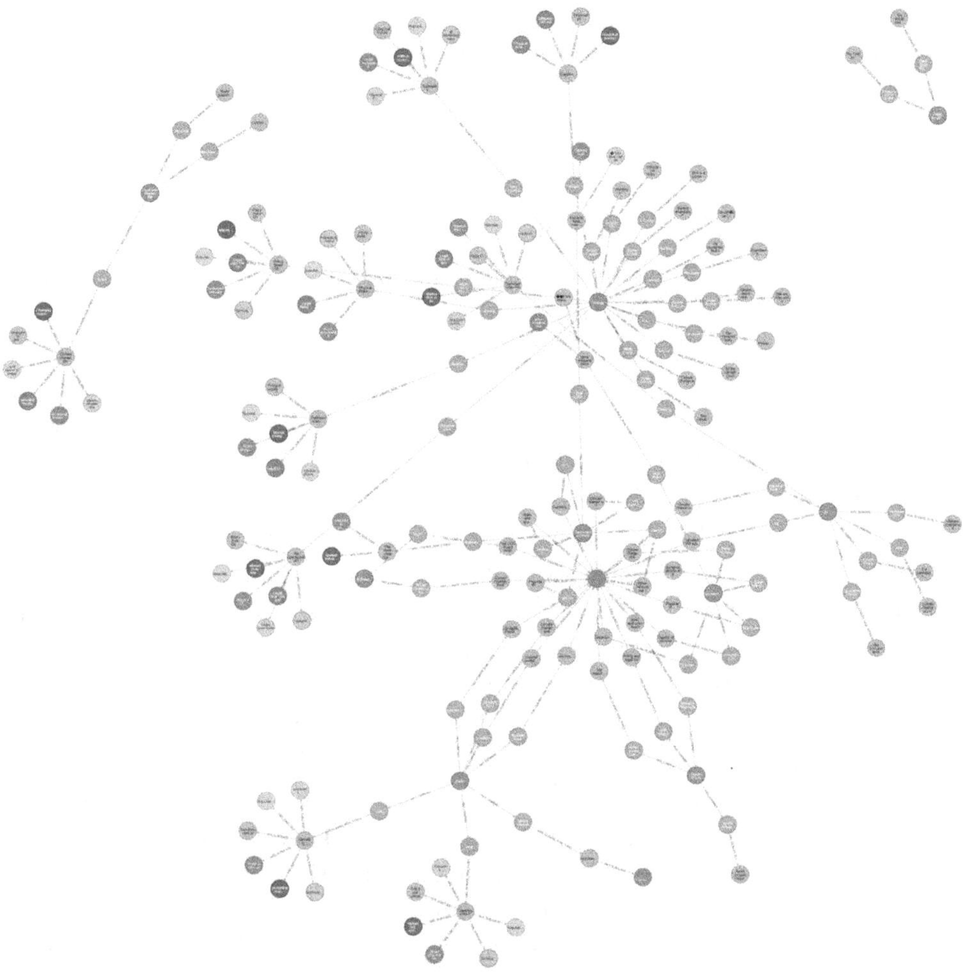

degree to which stakeholders are willing to accept risk in pursuit of certain outcomes, can differ greatly. Some may prefer conservative measures with low risk, while others might endorse more ambitious, riskier initiatives.

Consequently, the decision-making framework in this context must be robust and flexible enough to accommodate these diverse risk tolerances and uncertainties. This requirement underlines the necessity for an integrative risk-management approach that synergizes quantitative and qualitative methodologies.

Machine learning-enhanced text mining and resulting automatic data visualization indicates that Australian corporations primarily focus on immediate physical risks.

They have yet to concentrate to the same degree on long-term physical risks or transitional risks. This pioneering methodological approach highlights central themes primarily connected to the efforts necessary to handle immediate physical climate events. Furthermore, the analysis of visualisation results reveals that Australian corporations prefer risk reduction strategies for managing climate risk, requiring less change than risk avoidance approaches. This could suggest that Australian businesses do not view climate risks as necessitating a radical transformation in their operational practices. However, this hypothesis necessitates further, more comprehensive investigation.

This preliminary case study evaluating content across all reporting frameworks reveals that only a minor fraction of Australian companies choose to disclose their attempts to identify and manage the climate risks they face. Among these disclosing firms, approximately half adhere to the primary elements of the international industry-standard TCFD-reporting framework. The remaining entities utilize a variety of frameworks rooted primarily in environmental sustainability or financial reporting.

We recommend additional research on CRDs using both machine learning-assisted text mining and the resultant automatic data visualization frameworks. These investigations should be conducted using a temporal perspective to analyze climate risk themes over time and evaluate how climate risk practices and discourse evolve. A component of climate risk reporting unexplored in this analysis but worthy of further research involves assessing the metrics and targets Australian companies employ to outline their progress toward climate objectives. A temporal evaluation of the qualitative data disclosed in climate-risk reporting could potentially determine how, and to what extent, Australian firms are realizing their climate-related goals. Another significant study would be to compare the key climate-risk themes and practices in Australia with those in other countries, using the presented methodology to establish how Australia's CRDs compare on an international scale.

Ultimately, a wider risk-management approach that merges quantitative and qualitative-based methodologies may enable the analysis of a broader range of outcomes and also allow the uncertainty in their likelihoods to be investigated.

8. DISCUSSION

As presented above in *Section 7.* text mining encompasses a suite of methodologies and algorithms designed for the explorative discovery and representation of patterns and interconnections within textual data sets. This investigative approach is foundational to the analysis of textual data. Recent advancements in the field have been facilitated by the integration of diverse machine learning techniques into traditional text mining tasks and workflows.

Contrasting traditional text analysis, achieved via computer-assisted qualitative data analysis software (CAQDAS), which enables exhaustive knowledge discovery, text mining prompts questions about their interrelationship, particularly in relation to methodology and qualitative data analysis. Are these approaches rivals, or do they complement each other? What advantages does machine learning-assisted text mining bring to qualitative data analysis? What novel opportunities do these updated text mining procedures introduce, and which challenges do they address?

From a methodological viewpoint, machine learning-enhanced text mining is a versatile research tool applicable to both primary and secondary data analysis. Analytically, it necessitates the expertise and skills to amalgamate and thoroughly examine both qualitative and quantitative data. The latest advancements in computer science, especially in the domain of sophisticated statistical algorithms for machine learning, enable text mining solutions to enrich conventional exploratory analyses of qualitative data derived from documented expressions or centered around social actors' actions and events.

Such an approach not only facilitates a holistic understanding of the phenomena and processes underpinning qualitative research methodology but also allows for the development of analytical classification or predictive models founded on discovered patterns and regularities. Machine learning-enriched text mining employs analytical techniques and algorithms to identify patterns in the structure of textual data that would otherwise remain elusive due to human limitations in perception and time constraints.

Computer-aided analysis of qualitative data or text mining using machine learning techniques are innovative approaches in qualitative climate change research. Merging text mining enhanced by machine learning with CAQDAS methods may enhance the reliability of qualitative outcomes and raises the prestige of qualitative research. Hence, CAQDAS software can potentially offer a more comprehensive tool for analyzing textual data across linguistic, syntactic, semantic, and pragmatic dimensions.

At the methodological and analytical levels, machine learning-backed text mining and CAQDAS aren't as distinct as they might initially seem. The primary difference pertains to the level of automation in the data mining process, while the analytical procedures for knowledge discovery are similar.

Practically, text mining and CAQDAS diverge when it comes to the volume of textual data processed. While CAQDAS software typically works with smaller qualitative data sets, machine learning-supported text mining facilitates the analysis of data sets of virtually unlimited size.

As presented in the case study, text mining and machine learning algorithms enable fast and efficient viewing and analysis of information, performing calculations and analyses at a speed unattainable by traditional qualitative approaches. These methods' ability to manage vast volumes of textual data outperforms traditional qualitative methodologies due to their high-dimensionality constraints. Hence, text mining and machine learning analytical techniques and algorithms supplement and enhance not only the standard features of CAQDAS software but also the analytical and cognitive abilities of the researcher.

Unlike the *a priori* algorithms traditionally used in qualitative research, text mining supported by machine learning techniques discover relations between variables when there are no predetermined expectations or assumptions of their nature. Hypotheses are generated from data *a posteriori* rather than posited *a priori*. In a text mining and machine learning analysis process, many variables are considered and compared in many different combinations and configurations, and significant relations between them are sought.

Research based on text mining supported by machine learning techniques develops analytical models based on the configuration of data and variables, regardless of the size or type of the dataset and number of variables. In the construction of such models, machine learning-enhanced text mining is founded on inductive-deductive and inductive-abductive reasoning. Models represent empirical associations which, subsequently, are tested and interpreted. They usually result in in so-called "cognitive maps" showing the relations between documents, codes, categories, or concepts.

If one were to strictly follow Ann Lewins' definition of CAQDAS referring to the so-called qualitative analysis of qualitative data (Lewins and Silver 2007), one would rather exclude the application of text mining and machine learning techniques and algorithms in the field of computer-aided qualitative data analysis. However, a comparison of the analysis logics of text mining supplemented with machine learning and CAQDAS software shows that, on the analytical and methodological levels, they are epistemologically compatible rather than competitive.

Based on conducted preliminary case study (*Section 7.*), it can be concluded that as in many other methodological approaches to the analysis of qualitative data, text mining and machine learning 'encourages' the researcher to be open-minded when constructing models. In other words, text mining and machine learning techniques and algorithms fit perfectly into the various stages of computer-aided analysis of data in the knowledge discovery processes. This is notable, since machine learning-enhanced text mining in analytical practice can be applied to a variety of textual data types.

9. CONCLUSION

Machine learning-enhanced text mining enables methodological solutions that automate and improve the analysis of qualitative data. However, contrary to popular belief, text mining and machine learning analysis is not an automated activity devoid of human input. Like CAQDAS, it is an iterative process that requires researcher input.

In practice, text mining and machine learning methods are mostly semi-automatic (and often semi-supervised) and require deep knowledge of and familiarity with analytical techniques as well as domain expertise.

Working with CAQDAS programs teaches the researcher methodological rigour, adherence to procedures, and qualitative data analysis accuracy and precision, while text mining and machine learning unveils new patterns in the material, provides interdisciplinarity, and requires additional analytical and mathematical skills, which reflect positively on the quality of the analysis. Methodologically, machine learning-supported text mining is, in a sense, a reflection of the logic of grounded theory (Charmaz 2014). Analytically, it is similar to content analysis. However, while both approaches—CAQDAS and ML-enhanced text mining— use computer algorithms to analyse textual data, text mining supplemented with machine learning goes beyond that. It is characterised by its unique ability to process natural language and use the knowledge contained in subject and domain-specific vocabularies during the analysis process. Thus, the application of machine learning-enhanced text mining to qualitative research on climate change leads to a deeper understanding of the underlying processes reflected in text documents. It also enhances the integration of data from many different sources.

This data integration, on the other hand, leads to the systematic development and integration of knowledge and improves the quality of analysis and secondary qualitative research. For a climate change researcher, a qualitative one in particular, it is extremely important to approach the issue from the data side, to discover knowledge from that data, to construct high-dimensional analytical models of mechanisms and social actions and, consequently, to test relations and hypotheses based on variables in these models.

The expectation that knowledge is confined within data has always been a theme in qualitative research. Exploring large textual datasets using text mining and machine learning processing provides computer-aided qualitative data analysis techniques new and unprecedented capabilities to discover relations between sources, thereby broadening and deepening knowledge.

10. STATEMENTS AND DECLARATIONS

The authors declare that no funds, grants, or other support were received during the preparation of this manuscript.

10.1 Competing Interests

We have no conflicts of interest to disclose.

10.2 Data Availability

The datasets are publicly available in the ZENODO repository: https://doi.org/10.5281/zenodo.7233150

REFERENCES

Abualigah, L. M. Q., & Hanandeh, E. S. (2015). Applying genetic algorithms to information retrieval using vector space model. *International Journal of Computer Science, Engineering and Applications*, *5*(1), 19. doi:10.5121/ijcsea.2015.5102

Aggarwal, C. C., & Zhai, C. (2012). *Mining text data*. Springer Science & Business Media. doi:10.1007/978-1-4614-3223-4

APR, A. P. R. (2022). *Prudential practice guide: draft CPG 229 climate change financial risks*. APR.

Atwoli, L., Muhia, J., & Merali, Z. (2022). Mental health and climate change in Africa. *BJPsych International*, *19*(4), 86–89. doi:10.1192/bji.2022.14

Basu, A., Walters, C., & Shepherd, M. (2003). Support vector machines for text categorization. *36th Annual Hawaii International Conference on System Sciences, 2003*. IEEE. 10.1109/HICSS.2003.1174243

Bouaziz, S., Nabli, A., & Gargouri, F. (2019). Design a data warehouse schema from document-oriented database. *Procedia Computer Science*, *159*, 221–230. doi:10.1016/j.procs.2019.09.177

Buttcher, S., Clarke, C. L., & Cormack, G. V. (2016). *Information retrieval: Implementing and evaluating search engines*. Mit Press.

Caldwell, Z. (2020). Corporations and climate change: An investigation of mandatory climate risk disclosure in Australia. *Environmental and Planning Law*, *37*(1), 3–17.

Charmaz, K. (2014). *Constructing grounded theory*. Sage.

Chipalkatti, N., Le, Q., & Rishi, M. (2021). Sustainability and society: Do environmental, social, and governance factors matter for foreign direct investment? *Sage*, *14*(19), 6039.

Clark, L., Pantidi, N., Cooney, O., Doyle, P., Garaialde, D., Edwards, J., Spillane, B., Gilmartin, E., Murad, C., & Munteanu, C. (2019). What makes a good conversation? Challenges in designing truly conversational agents. *Proceedings of the 2019 CHI Conference on Human Factors in Computing Systems*. ACM. 10.1145/3290605.3300705

Don, A., Zheleva, E., Gregory, M., Tarkan, S., Auvil, L., Clement, T., Shneiderman, B., & Plaisant, C. (2007). Discovering interesting usage patterns in text collections: integrating text mining with visualization. *Proceedings of the sixteenth ACM conference on Conference on information and knowledge management*. ACM. 10.1145/1321440.1321473

Dumais, S. T. (2004). Latent semantic analysis. *Annual Review of Information Science & Technology*, *38*(1), 188–230. doi:10.1002/aris.1440380105

Erk, K., & Padó, S. (2008). A structured vector space model for word meaning in context. *Proceedings of the 2008 Conference on Empirical Methods in Natural Language Processing*. ACM. 10.3115/1613715.1613831

Fan, W., Wallace, L., Rich, S., & Zhang, Z. (2006). Tapping the power of text mining. *Communications of the ACM*, *49*(9), 76–82. doi:10.1145/1151030.1151032

Gartner, T. (2008). *Kernels for structured data*. World Scientific. doi:10.1142/6855

Grishman, R. (2019). Twenty-five years of information extraction. *Natural Language Engineering*, *25*(6), 677–692. doi:10.1017/S1351324919000512

Gruber, T. R. (1993). A translation approach to portable ontology specifications. *Knowledge Acquisition*, *5*(2), 199–220. doi:10.1006/knac.1993.1008

Guo, J., Fan, Y., Pang, L., Yang, L., Ai, Q., Zamani, H., Wu, C., Croft, W. B., & Cheng, X. (2020). A deep look into neural ranking models for information retrieval. *Information Processing & Management*, *57*(6), 102067. doi:10.1016/j.ipm.2019.102067

Gupta, V., & Lehal, G. S. (2009). A survey of text mining techniques and applications. *Journal of Emerging Technologies in Web Intelligence*, *1*(1), 60–76. doi:10.4304/jetwi.1.1.60-76

Han, X., Gao, T., Yao, Y., Ye, D., Liu, Z. & Sun, M. (2019). *OpenNRE: An open and extensible toolkit for neural relation extraction.*

Honnibal, M. & Montani, I. (2017). spaCy 2: Natural language understanding with Bloom embeddings, convolutional neural networks and incremental parsing. *Sage*, *7*(1), 411-420.

Jo, T. (2019). Text mining. *Studies in Big Data.*

Kadhim, A. I. (2019). Survey on supervised machine learning techniques for automatic text classification. *Artificial Intelligence Review*, *52*(1), 273–292. doi:10.100710462-018-09677-1

Kannan, S., V. Gurusamy, S. Vijayarani, J. Ilamathi, M. Nithya, S. Kannan, V. J. I. J. o. C. S. Gurusamy and C. Networks (2014). Preprocessing techniques for text mining. *Sage, 5*(1), 7-16.

Korhonen, A., Ó Séaghdha, D., Silins, I., Sun, L., Högberg, J., & Stenius, U. (2012). Text mining for literature review and knowledge discovery in cancer risk assessment and research. *PLoS One*, *7*(4), e33427. doi:10.1371/journal.pone.0033427 PMID:22511921

Kowsari, K., Brown, D. E., Heidarysafa, M., Meimandi, K. J., Gerber, M. S., & Barnes, L. E. (2017). Hdltex: Hierarchical deep learning for text classification. *2017 16th IEEE international conference on machine learning and applications (ICMLA)*. IEEE. 10.1109/ICMLA.2017.0-134

Kumpu, V. (2022). What is public engagement and how does it help to address climate change? A review of climate communication research. *Environmental Communication*, *16*(3), 304–316. doi:10.1080/17524032.2022.2055601

Lal, M. (2015). *Neo4j graph data modeling*. Packt Publishing Ltd.

Lewins, A., & Silver, C. (2007). Qualitative coding in software: Principles and processes. *Using Software in Qualitative Research.*, *9780857025012*.

Luccioni, A., & Palacios, H. (2019). Using natural language processing to analyze financial climate disclosures. *Proceedings of the 36th International Conference on Machine Learning*, Long Beach, California.

Mannai, M., Karâa, W. B. A., & Ghezala, H. H. B. (2018). Information extraction approaches: A survey. Information and Communication Technology. Springer.

McGilvray, J. (2014). Chomsky: Language, mind and politics. *Polity*.

Minaee, S., Kalchbrenner, N., Cambria, E., Nikzad, N., Chenaghlu, M., & Gao, J. (2021). Deep Learning—based Text Classification: A Comprehensive Review. *ACM Computing Surveys*, *54*(3), 1–40. doi:10.1145/3439726

Mitra, B., & Craswell, N. (2017). Neural models for information retrieval. arXiv preprint arXiv:1705.01509.

Moens, M.-F. (2006). *Information extraction: algorithms and prospects in a retrieval context*. Springer Science & Business Media.

Moser, S. C. (2016). Reflections on climate change communication research and practice in the second decade of the 21st century: What more is there to say? *Wiley Interdisciplinary Reviews: Climate Change*, *7*(3), 345–369. doi:10.1002/wcc.403

Ogheneovo, E., & Japheth, R. (2016). Application of vector space model to query ranking and information retrieval. *International Journal of Advanced Research in Computer Science and Software Engineering*, *6*(5).

Pearce, W., Grundmann, R., Hulme, M., Raman, S., Hadley Kershaw, E., & Tsouvalis, J. (2017). Beyond counting climate consensus. *Environmental Communication*, *11*(6), 723–730. doi:10.1080/17524032.2017.1333965

Porter, M. F. (1980). *An algorithm for suffix stripping*. Program. doi:10.1108/eb046814

Pujara, J., & Singh, S. (2018). Mining knowledge graphs from text. *Proceedings of the Eleventh ACM International Conference on Web Search and Data Mining.* ACM. 10.1145/3159652.3162011

Rolnick, D., Donti, P., Kaack, L., Kochanski, K., Lacoste, A., Sankaran, K., Ross, A., Milojevic-Dupont, N., Jaques, N., & Waldman-Brown, A. (2022). Tackling climate change with machine learning. *Sage, 55*(2), 1-96.

Roshdi, A., & Roohparvar, A. (2015). Information retrieval techniques and applications. *International Journal of Computer Networks and Communications Security*, *3*(9), 373–377.

Rybak, N., & Hassall, M. (2021). Deep Learning Unsupervised Text-Based Detection of Anomalies in US Chemical Safety and Hazard Investigation Board Reports. *2021 International Conference on Electrical, Computer, Communications and Mechatronics Engineering (ICECCME).* IEEE. 10.1109/ICECCME52200.2021.9590834

Sabban, I., Lopez, C., & Mercuzot, Y. (2020). Automatic analysis of insurance reports through deep neural networks to identify severe claims. *Annals of Actuarial Science*, 1-26.

Salton, G., Wong, A., & Yang, C.-S. (1975). A vector space model for automatic indexing. *Communications of the ACM*, *18*(11), 613–620. doi:10.1145/361219.361220

Sanderson, H. (2019). How do climate risks affect corporations and how could they address these risks? *Sage, 1,* 1-6.

Schütze, H., Manning, C. D., & Raghavan, P. (2008). *Introduction to information retrieval*. Cambridge University Press Cambridge.

Shafiabady, N., Lee, L. H., Rajkumar, R., Kallimani, V., Akram, N. A., & Isa, D. (2016). Using unsupervised clustering approach to train the Support Vector Machine for text classification. *Neurocomputing*, *211*, 4–10. doi:10.1016/j.neucom.2015.10.137

Sperlich, W. B. (2006). *Noam Chomsky*. Reaktion Books.

Sun, W., Cai, Z., Li, Y., Liu, F., Fang, S., & Wang, G. (2018). Data processing and text mining technologies on electronic medical records: A review. *Journal of Healthcare Engineering*, *2018*, 2018. doi:10.1155/2018/4302425 PMID:29849998

Toivonen, H. (2022). Themes of climate change agency: A qualitative study on how people construct agency in relation to climate change. *Humanities & Social Sciences Communications*, *9*(1), 102. doi:10.105741599-022-01111-w

Turing, A. M., & Haugeland, J. (1950). *Computing machinery and intelligence*. MIT Press Cambridge. doi:10.1093/mind/LIX.236.433

Verma, S., & Nidhi, V. (2017). *Extractive summarization using deep learning*. arXiv preprint arXiv:1708.04439.

Vijayarani, S., Ilamathi, M., & Nithya, C.(2015). Preprocessing techniques for text mining-an overview. *Sage, 5*(1), 7-16.

Weiss, S. M., Indurkhya, N., & Zhang, T. (2015). *Fundamentals of predictive text mining*. Springer. doi:10.1007/978-1-4471-6750-1

Weizenbaum, J. (1966). ELIZA—A computer program for the study of natural language communication between man and machine. *Communications of the ACM*, *9*(1), 36–45. doi:10.1145/365153.365168

Wong, P. C., Whitney, P., & Thomas, J. (1999). Visualizing association rules for text mining. *Proceedings 1999 IEEE Symposium on Information Visualization (InfoVis' 99)*. IEEE.

Yang, Y., & Liu, X. (1999). A re-examination of text categorization methods. *Proceedings of the 22nd annual international ACM SIGIR conference on Research and development in information retrieval*. ACM. 10.1145/312624.312647

Zaccai, E. (2012). Over two decades in pursuit of sustainable development: Influence, transformations, limits. *Environmental Development*, *1*(1), 79–90. doi:10.1016/j.envdev.2011.11.002

Zhai, C., & Lafferty, J. (2017). A study of smoothing methods for language models applied to ad hoc information retrieval. *ACM SIGIR Forum*. ACM New York, NY, USA. 10.1145/3130348.3130377

ADDITIONAL READING

Bishop, C. M., & Nasrabadi, N. M. (2006). *No. 4* (Vol. 4). Pattern recognition and machine learning. Springer.

Goodfellow, I., Bengio, Y., & Courville, A. (2016). *Deep learning*. MIT press.

Harker, C., Hassall, M., Lant, P., Rybak, N., & Dargusch, P. (2022). What Can Machine Learning Teach Us about Australian Climate Risk Disclosures? *Sustainability (Basel)*, *14*(16), 10000. doi:10.3390u141610000

Harrington, P. (2012). *Machine learning in action*. Simon and Schuster.

Khan, A., Baharudin, B., Lee, L. H., & Khan, K. (2010). A review of machine learning algorithms for text-documents classification. *Journal of advances in information technology*, *1*(1), 4-20.

Miner, G., Elder, I. V. J., Fast, A., Hill, T., Nisbet, R., & Delen, D. (2012). *Practical text mining and statistical analysis for non-structured text data applications*. Academic Press.

Rybak, N., & Hassall, M. (2022). Machine Learning–Enhanced Decision-Making. Handbook of Smart Materials, Technologies, and Devices: Applications of Industry 4.0, 477. doi:10.1007/978-3-030-84205-5_20

Weiss, S. M., Indurkhya, N., & Zhang, T. (2015). *Fundamentals of predictive text mining*. Springer. doi:10.1007/978-1-4471-6750-1

KEY TERMS AND DEFINITIONS

Algorithmics: A subfield of computer science. Study of design and efficiency of algorithms.

Deep Learning: Deep learning is a subfield of machine learning. Deep neural networks (DNNs) are the extension of artificial neural networks (ANNs). Deep learning methods and techniques scale up the size and complexity of ANNs to produce increasingly richer functionality and very often archive better results in disciplines such as text mining or computer vision.

Entropy: In information theory, average amount of information attributable to a single message from information source.

Information Extraction (IE): IE is the process of identifying and extracting content from documents written in natural language and is based on analytically generated or predefined knowledge patterns

Information Retrieval (IR): IR is the process of searching for and locating textual information in a database based on a user query. Searches can be based on a document's existing metadata, its full text or its content indexing.

Labelled Data: In machine learning, data that have already been categorized.

Lemmatisation: Morphological analysis of the dictionary and reduction of similar lexical forms of words to one basic form – a *lemma*. A lemma is the canonical, simplest form of a lexeme used for its dictionary representation.

Prediction: In machine learning, calculating the value of output variable for input variable values falling within training data values.

Probability Distribution: Function used to compute the likelihoods of occurrence of various observation outcomes.

Statistics: In the context of text mining, 'statistics' refers to the quantitative methods used to analyze and interpret text data. This can involve various techniques such as frequency analysis (counting the occurrence of words or phrases), topic modeling (identifying the main topics in a text), and association rules (finding relationships between words or phrases). These statistical methods help to transform unstructured text data.

Tokenization: Tokenization of a text document is the division of the input text into sentences, words, punctuation marks (commas, full stops, etc.) and non-text characters. Other elements of a document that have the status of regular expressions can also be used as tokens. Tokenization is usually an automatic process, depending only on the formal language of the text mining programme.

Chapter 5
Social Media and Gen Y at Work:
The Uses and Gratifications of Technology

Pallavi Ghanshyala Vyas
https://orcid.org/0000-0002-1017-0308
Jain University, India

Swarn Priya
Symbiosis International University (Deemed), India

ABSTRACT

Workplaces have been revolutionized by two factors: the rise of the Millennial age into the workforce and the emergence of online social networking tools as a means of connecting and collaborating with others. With the increasing penetration of technology in the form of artificial intelligence, IOT, 5G networks, and varied forms of social media, it is opportune for organizations to understand their impact on the current workforce. This research examines how Gen Y (millennials) use social media at work and their consequences on firms. Millennial employees seek organizational benefits such as increased knowledge sharing, work-related learning, employee engagement, and bridging social capital. The antecedents and consequences of social media use by Gen Y employees have been explained according to the uses and gratifications (U&G) theory.

DOI: 10.4018/978-1-6684-8634-4.ch005

INTRODUCTION

There is a rising concern that smart machines will soon replace many humans in decision-making, as artificial intelligence (AI) has already permeated many organizational processes. IBM's Watson and Google DeepMind's AlphaGo are prime examples of the recent prominence and rapid progress in artificial intelligence. The field of AI encompasses a diverse array of applications and methodologies, including but not limited to neural networks, speech and pattern recognition, genetic algorithms, and deep learning. The application of AI is embedded in various social media platforms that detect users' behaviors and patterns and gives insights to various companies on how to engage with their customers. Web 3.0 technology and its uses are becoming increasingly embedded in people's daily lives, especially among young people from a variety of backgrounds and communities. The proliferation of cellular networks and internet-enabled mobile devices has had a significant impact on how individuals see the world around them. At a time when the globe seems to be in the midst of a social media revolution, it's clear that people of all ages and backgrounds are turning to online platforms like Facebook, Twitter, and Instagram to have meaningful conversations with one another. Digitalization is rapidly altering the primary source of experiences that people have and is redefining the association between members of a generation and the community of others (Au, 2020). Furthermore, it is often believed that members of the younger generations (such as Gen Y) are avid users of digital media particularly, social networking sites. The demographics have shifted due to Baby Boomers retiring and Millennials (or Gen Y) and Gen Z entering the workforce. Researchers recommend adapting 2020 workplaces to the unprecedented challenge of five generations of employees working together, emphasizing the importance of digital revolution and mobile phones for communication and learning (Meister & Willyerd, 2010). These Gen Y are seen as the future of the company and a driving force in the workforce (Kaifi, Nafei, Khanfar, & Kaifi, 2012). For members of Generation Y, connection is a key driver in their social media usage (Berezan, Krishen, Agarwal, & Kachroo, 2018). Companies must provide incentives that appeal to the Xers, Yers, and Zers that make up the bulk of today's workforce before attracting the best candidates of each generation for employment (Mahmoud et al., 2021) as it may additionally pose innumerable issues for managers while trying to manage people dynamics, as by 2025, Millennials will make up 75% of the world's labor force (Cristea et al., 2017). Also, a quarter of the Asia-Pacific region's population will be made up of Gen Zers by 2025, and as they reach adulthood, they will have greater purchasing power. While they do share many characteristics with millennials, they are also uniquely their own (Kim, Mclnerney, & Smith, 2020). While still coming into their own, members of Generation Z are accelerating millennial-era developments like the search for authenticity and the use of social media. Although Gen Zers share millennials' enthusiasm for social media, they differ from millennials in that they don't blindly follow the crowd; instead, they look for sources they trust (Bhargava, Finneman, Schmidt, & Spagnuolo, 2020). While Generation Y is highly tech-savvy, reaching them requires providing them with convenient digital options (Weyland, 2011). For both employee engagement and retention, social media has been acknowledged as a critical factor influencing employees' developmental characteristics (Bolton, Parasuraman, Hoefnagels, Migchels, et al., 2013).

Social networking is a popular Internet activity today. As of January 2023, Statista estimated that 5.16 billion people, or 64.4% of the world's population, were online. This means that 5.94% of the world's population, or 4.76 billion individuals, regularly engage in social media. In 2022, Asia had more internet users than any other region, with over 2.9 billion. With over 750 million people, Europe is in second place for internet penetration. When compared to the rest of the globe, India, US and China have extremely high internet penetration rates (Meister & Willyerd, 2010). Social networking, one of the most extensive activities on the Internet, has played a major role in facilitating online cooperation in organizations (Franchi et al., 2016), which has contributed to the emergence of a communication revolution over the past few decades (Massari, 2009). Social Networks can be defined as "applications, services and systems that allow users to create, remix and share content" (Junco, 2014). Due to an ease of accessing internet, the number of social media users in India was 326.1 million in 2018, with the number expected to increase to almost 448 million in 2023 (Keelery, 2020). Facebook is the most popular social networking site with 1.9 billion unique monthly users out of which 83% are predominantly females in contrast to 76% of online men (Hutchinson, 2017). Much of this Facebook usage comes from the younger generation, between 18- 24 years of age, with around 97 million users in 2018 (Keelery, 2020) and it's said to be the best place to reach Millennials and Gen X (ages 18- 49 years), with Gen X spending almost 7 hours per week on social media (Hutchinson, 2017). Indian internet users increased due to digitalization and low data pricing. 467 million were active social media users. In the third quarter of 2022, WhatsApp and Instagram had the highest penetration (*India: Social Media Penetration 2022 | Statista*, n.d.). Brunei had 116.5 percent of the Asia-Pacific region's active social media users as of February 2022. The next closest country, with a penetration rate of 91.7%, was Malaysia. When compared, Indonesia's rate of social media adoption was significantly higher, at 68.9 percent. About 78.5% of Malaysians were regular users of social media as of January 2023. From 2022, when 91.7% of Malaysians were estimated to be active social media users, we can see a decline of 13.2%. Facebook dominated the other social media options that were readily available (*APAC: Active Social Media Penetration by Country 2022 | Statista*, n.d.).

Users these days rely more on social media for various reasons like entertainment, web searches, news and others (Giannakos, Chorianopoulos, Giotopoulos, & Vlamos, 2013). The current drift towards social media can truly be related with how the internet evolved and the way it retransformed the World Wide Web to its sole reason for being created, i.e., as a platform to smooth the progress of information exchange between its users (Kaplan & Haenlein, 2010). Social media use is basically for collaborating, participation and sharing (Kaplan & Haenlein, 2010). It provides users with the affordances of visibility, persistence, editability and association thus providing its users the flexibility in using the communication technology the way they want to (Treem & Leonardi, 2012).

Using SNSs at work can have numerous positive effects, such as better strategic clarity, streamlined operations, enhanced innovation, heightened employee involvement, and higher productivity (Bennett et al., 2010). Social networking sites (such as Facebook) are commonly used to keep in touch with friends and family. After the bulk of one's offline connections have been made, the majority of one's time spent on these sites is spent maintaining those connections (Lampe, Ellison, & Steinfield, 2008). By enhancing employees' ability to communicate and collaborate, social networking technology helps businesses become more proactive and responsive to their customers' needs. It also provides tools for creating, maintaining, and stimulating individuals' own social networks (Bennett, Pitt, & Owers, 2010). Remote workers can feel more connected to the company's culture and more engaged in their job thanks to the availability of social networking platforms. It allows users to interact with clients and customers, which improves communication and the sharing of information. (Bell, Graham, Hardy, Harrison, Stansall, & White, 2008; Bennett, et al., 2010). However, further research is needed to determine how SNS use correlates with other outcomes in the workplace (Nkwe & Cohen, 2017). Studies have found demographics to have an impact on social networking sites use. One study noted generational differences in the use of social media platforms like Facebook, LinkedIn, and Twitter, finding that the youngest cohort was increasingly active on Facebook and that a surprisingly high percentage of LinkedIn users were aged 25 to 45 (Archambault & Grudin, 2012). According to the results of numerous studies, there are generational differences that can be uncovered by looking at the effect of organizational variables such as industry, organization size, organizational culture and structure, and individual variables such as gender, sexual orientation, socioeconomic status, tenure, family status, etc. (Joshi, Dencker, Franz, & Martocchio, 2010; Laufer & Bengtson, 1974; Parry & Urwin, 2011). Studies on the age and social relationship dimensions indicate that young adults have a propensity to be future-oriented to a large extent, and thus place greater importance on extension of their network, while older people prefer to engage more in emotional interface with their close ties (Lang, 2001). Also, digital media and social networking sites have resulted in a transformation of cultural norms and processes within social lives of people to level the lived experiences of the generations (Au, 2020). These generational differences have been a very frequently discussed topic among researchers with numerous theoretical evidences of studies comparing generations with respect to various personal and work- related variables like, their personalities, teamwork and leadership preferences, work values and attitudes, career experiences and leader behaviors. With the projections of an aging workforce, the organizations will have to shift their focus to the Gen Y as one of its crucial workforces.

THEORETICAL FRAMEWORK

Workplace Social Media Use by Gen Y

According to generational cohort theory, a generation is a group of people who share similar childhood experiences and share values, viewpoints, and attitudes (Kupperschmidt, 2000). Generation Y is defined by early technological exposure, which has both beneficial and negative consequences on mental and emotional development and social connections (Immordino-yang et al., 2012). In its broadest meaning, the word "social media" refers to any service that allows its users to produce and disseminate content. In spite of the fact that social media have been around since before the birth of Generation Y (1981), widespread use did not occur until around 2003 (Boyd & Ellison, 2008). Generation Y in wealthy nations like the US uses technology as their "third hand" and "second brain"(Khal et al., 2013). As per studies, people aged 65 and up were found to rapidly increase their digital literacy and proficiency, and adapted to the same SNS digital culture and habits as younger generations Quan-Haase, et al., 2018). The Millennials or Gen Y are known to share content online without analyzing the information in advance (Haydam, Purcarea, Edu, & Negricea, 2017) in contrast to Gen Xers which are said to act more sensibly in their use of social media communication as they got familiarized with the modern communication technology and social media after reaching adulthood (Dabija & Grant, 2016). As suggested by several studies people from older generations can be found to be quite comfortable with novel technologies while those from the digital native generation can be seen to be restricting themselves to only a few (Thompson, 2012). Even among the same age groups in the digital native generations few of them restricted themselves to only using essential technology features, like using cell phones and basic web use like sending e-mails or looking for information, while, still others use technology for gaming, multimedia content creation or other such smartphone features (Kennedy, et al., 2010). Through their study Vitak, Crouse, & LaRose, (2011), reported that being younger positively predicted social media use whereas, another study by Andreassen, Torsheim, & Pallesen, (2014) reported that the use of social networking sites like Facebook or Twitter during work hours was negatively related to the age of the users.

The early communication theories like 'Uses and Gratifications (U&G) Theory' has been used to examine social media use by individuals. Users actively choose and use media in order to fulfil certain specific needs and thus obtain gratifications from the media. As for example, if an individual has a need for escape, there are particular media available to gratify the need satisfactorily (Quan-haase & Young, 2010). The 'why' and 'how' of media consumption are often referred to as the 'uses and gratifications. It's a term for discussing the reasons behind and the pleasure gained from specific uses. These pleasures can be broken down into two categories: content gratifications, which are based on what's in the media, and process gratifications, which are based on what happens when you use the media. Social gratification has been identified as a third form of internet-related satisfaction as internet use increases (Stafford, Stafford, & Schkade, 2004). The uses and gratifications approach emphasizes individual worth through examining how people utilize media (friend-networking sites). College students used friend-networking sites to "keep in touch with old and current friends, to post/look at pictures, make new friends, locate old friends," according to research (Raacke & Bonds- Raacke, 2008). The U&G approach explains the perceived usefulness of media using imperative and consequential concepts. The psychological and social context, the requirements and motivations for using media for communication, perspectives on and assumptions about media, viable alternatives to media consumption, communication patterns, and the outcomes and consequences of such actions all play a role (Kwon et al., 2013). According to the "Uses and Gratifications" paradigm, social networking sites provide traditional content gratification, social capital building, social networking surfing, surveillance, and communication. However, not all users utilize social media for the same reasons or get the same benefits. Social gratifications increase frequency, while content gratifications increase SNS time. SNS users use Facebook for a variety of purposes, which challenges site designers (Joinson, 2008). According to research, there are several benefits to utilizing social media, including gaining information, spending time with friends, being entertained, sharing one's opinions, relieving stress, and learning more about other people (Whiting & Williams, 2013).

Social media can change and reaffirm life events, positively and unfavorably (Scheinbaum, 2017). Social media can aid open-ended knowledge management (Nkwe & Cohen, 2017). It provides an environment for self-initiated learning whereby participants establish a framework of knowledge and assistance by creating and nurturing existing relationships (Yan et al., 2010) as well as people with self-regulation capacity communicate with peers to receive feedback and improve performance (Wang & Wu, 2008). Wikis for group collaboration, blogs for learning, podcasting and videos for research, and other channels for discussion and idea sharing assist many organizations (Shepherd, 2011). A vast amount of intentional and unintentional learning is expedited every now and then owing to the posts on Facebook, Twitter, LinkedIn, YouTube, Pinterest, and numerous other social media services. A greater consumption of internal social media results in greater self- reported employee engagement (Haddud et al., 2016). Technology advancements like the Internet allow for increased communication, collaboration, and knowledge sharing, all of which contribute to greater job satisfaction (Castellacci & Viñas-bardolet, 2019). The utilization of social media platforms has been found to contribute to the improvement of employees' effectiveness in the workplace through facilitating the exchange of knowledge. Nevertheless, it does not exert any influence on their bridging social capital (Vyas & Itam, 2022).

Antecedents of Gen Y's Social Media Use

In light of the proliferation of social media, it is crucial to investigate the factors that motivate users and the effects that their actions have on their happiness and ability to connect with others. A consumer's social media use can be influenced by peer pressure, but not by the consumer's own core beliefs. Furthermore, consumers' subjective well-being and social capital are both enhanced by their use of social media. The empirical research also reveals that there are no substantial differences in social media usage based on consumer demographic factors (age and gender) (Karikari, Osei-Frimpong, & Owusu-Frimpong, 2017). It is impossible to overstate the impact that social media has had and will have on people's lives and businesses around the world (Ngai et al., 2015). Consumers' social media use is influenced by a number of antecedents, many of which have their consequences (Bolton, Parasuraman, Hoefnagels, Migchels, et al., 2013). The social presence theory has been used in the past to examine the antecedents and consequences of social media use. According to the social presence theory, the way in which a medium is used to convey information from one person to another can affect how well the information is received and processed by the recipient (Cui et al., 2013). This boosts the user's confidence in engaging in social interactions (Lowenthal), which is in turn influenced by the user's own values and the norms of his or her social referent groups (Karikari, Osei-Frimpong, & Owusu-Frimpong, 2017). Environmental and individual variables as well as economic, technological, cultural, and political/legal aspects affect social media use. Individual variances originate from generally stable elements like socioeconomic position, personal values/preferences, age/lifecycle stage, and dynamic factors like objectives, emotions, and social norms that may vary during social media use (Bolton, Parasuraman, Hoefnagels, Migchels, et al., 2013).

Thus, based on the previous research the following antecedents and consequences of social media use by Gen Y employees have been uncovered:

***Economic Environment*:** Since a country's economic climate affects people's disposable income, job security, trust in the economy, etc., it may cause them to alter their social media behavior. Consumer spending, notably on hardware that allows access to social media, may decline as a result of tighter budgets brought on by an economic downturn (Kreutzer, 2009). Affordability is a crucial indication of social media use adoption since it represents the desire to pay for gadgets and services across countries and marketplaces. Internet connectivity is essential to narrowing the digital gap with regards to and countryside and metropolitan areas (Hargittai & Hinnant, 2008) and high and low income economies (Andrés & Cuberes, 2014).

***Culture*:** Cultures perceive, accept, and use communication and information technologies differently (Srite & Karahanna, 2006; Van Slyke et al., 2005). Repeated player interactions build trust, which is favorably connected with social media penetration. Every member of Generation Y uses social media for diverse reasons, having distinct feelings, and according to numerous social and cultural expectations and identities (Bagozzi & Dholakia, 2002).

***Technology*:** Media technology includes many devices, channels, and places, such as social networking sites, where users can engage with each other (Sundar & Limperos, 2013). Relationships, both new and old, can be fostered or preserved with the aid of technology (Rossmann et al., 2016). Technology's key roles are to improve search and exchange of information (its functional utility), hedonic utility (pleasure) and social utility (information exchange and engagement) among its users (Shao & Kwon, 2018).

***Generational Identity*:** Individuals' levels of technology use and engagement may be influenced by the generational cohort in which they were raised, as there are significant differences in how each generational group uses technology (Calvo-Porral & Pesqueira-Sanchez, 2020; Magsamen-Conrad et al., 2015). Technology dominates millennials' lives, shaping their expectations and worldview (Taylor & Keeter, 2010).

***Individual Characteristics*:** Socioeconomic background, individual principles, and maturity phase also influence millennial's media platforms use (Bolton, Parasuraman, Hoefnagels, Kabadayi, et al., 2013). Those with less formal education may be less likely to use the internet effectively and more interested in browsing for amusement than learning (Hargittai & Hinnant, 2008). Facebook is mostly used to preserve social ties and reconnect with old pals (Ellison, 2007). It helps students' network and learn about social activities (Lampe et al., 2007); (Raacke et al., 2008). The users also indicate a 'Wasting time' motivation behind Facebook use (Giannakos et al., 2013). Further benefits of using Facebook include social inclusion (i.e., being part of the space where peers communicate) and keeping connected with offline peers (Quan-haase & Young, 2010). Social media is used for information searching, spend time, social engagement, entertaining, expression of opinion, expediency utility, relaxation, communicatory utility, and knowledge about others (Whiting et al., 2013).

***Organizational Context*:** Web-based technologies impact HRM system efficacy (Huub et al., 2023; Strohmeier, 2007). IT is crucial to a company since information coordinates value chain activities and can introduce new competitive advantages and business. Therefore, organizational structure and communication investments must accompany IT investments to achieve these aims (Melville et al., 2004). Media with a larger social presence are chosen over ambiguous and uncertain methods of communication because they can produce direct or indirect human contact and make communication rapid, initiating a dialog process similar to in-person (Gefen & Straub, 2004; Cheung et al., 2011).

Outcomes of Social Media Use at Work

***Work-Related Learning*:** Bandura's social learning theory states that people interact with peers and surroundings to learn. Self-directed learning leads to desired learning outcomes (Bandura, 1977). Learning is facilitated by social media because it facilitates communication within a community and the dissemination of content produced by its users (Kim et al., 2010). A study of nurses found that the more they utilized social media at work, the more likely they were to take part in learning opportunities (van Puijenbroek et al., 2014). Learning is facilitated by social media because it facilitates communication within a community and the dissemination of content produced by its users (Kim et al., 2010). Workers might use social media to connect with other experts in their field or to promote their own knowledge (Shepherd, 2011). Learning in the workplace can benefit from the use of social media as a supplement to other learning activities (van Puijenbroek et al., 2014).

***Bridging Social Capital*:** Bridging social capital comes from "weak ties"—loose relationships between people who share new viewpoints and helpful knowledge but not emotional support, which can lead to employment opportunities (Granovetter, 1983). According to their research (Ellison, 2007), discovered that heavy Facebook use strongly predicted the social connections between people. The benefits that people gain from their networks of loved ones, coworkers, and acquaintances are highlighted by the term "bridging social capital" (Franchi et al., 2016). People invest in social relations in anticipation of certain gains (Pribeanu & Balog, 2017) and social networking sites are linked with individual's bridging social capital (Horng & Wu, 2020). Even for rural communities, greater levels of SNS use are deemed helpful in creating and sustaining bridging social capital online (Tiwari et al., 2019).

***Knowledge Sharing*:** Recent works on knowledge management have highlighted the value of Web 2.0 technologies, which include blogs, wikis, and other forms of social media, for bringing a more human element to the process of knowledge management as a whole (Ardichvili et al., 2003). The use of SNSs in the workplace has advantages like higher levels of collective knowledge, better understanding, and more output (Bennett et al., 2010). Employees contribute their knowledge to the Web 2.0 platform for communicating effectively, organizing individual knowledge, creating dialogue about novel ideas and concepts, finding solutions to specific problems, remain up-to-date on the recent headlines and actions of co workers, obtaining wanted assistance and suggestions, and expanding their social media network (Paroutis & Saleh, 2009). Organizational learning and development depend on the widespread adoption of social media in the workplace (Ellison et al., 2014). Some of the most important conditions for conveying tacit information via social media can be met by these platforms (Ma & Chan, 2014). Knowledge sharing can be predicted by a person's traits, the amount of knowledge created online, and the confidence the person has in their own abilities (Kim et al., 2019). Social networking sites use at the worklace has been associated ositively with knowledge sharing (P.G. Vyas & Pandey, 2020).

***Employee Engagement*:** Organizational engagement is defined as active participation in achieving the organization's mission (Macey, 2009). With the help of social media apps, professionals may easily and casually network with one another and exchange both personal and professional updates (Haddud et al., 2016). Internal social media can reinforce employee engagement, making it a sort of "blended" employee engagement (Sievert & Scholz, 2017). An indistinctive social media use may help survive organizational change and may Indistinct social media use may help survive organizational transformation and affect engagement outcomes (Heewon Kim & Scott, 2019).

The relationships between the antecedents and consequences of SM use have been represented in figure 1.

Figure 1. Antecedents and consequences of social media (SM) use

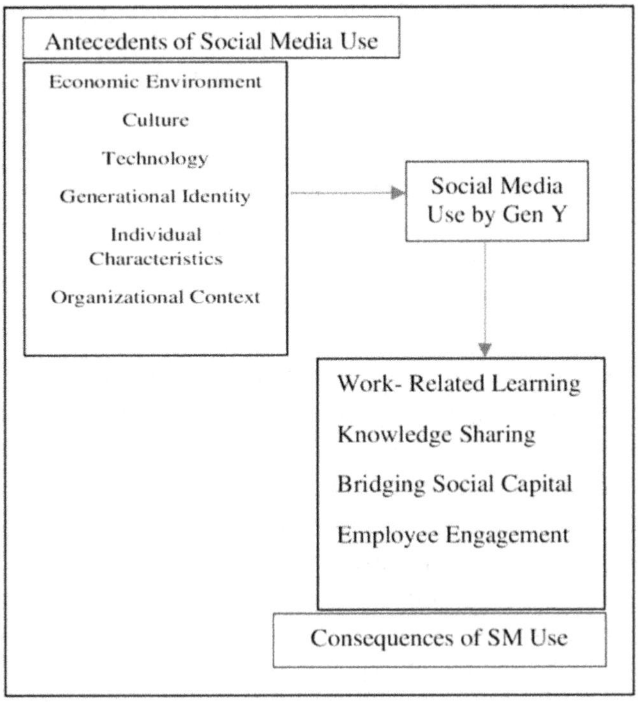

DISCUSSION

Researchers worldwide have studied generational variety for 20 years due to significant changes in demographics across the world. A number of Western countries have been experiencing the veracity of ageing whereas countries like India have a demographic surplus, with the average age being 25 years (Srinivasan, 2012). The concept of generations comprises a combination of social relations and processes that need to be interrogated due to the changing concepts of time and speed due to digitalization. The absolute penetration of SNSs in terms of demographics or technology has not yet arrived, though digitalization does bring people closer (Au, 2020).

Existing research on social media use shows that businesses are underestimating the benefits and profitability of social networking in the workplace. These commercial benefits outweigh pre-millennials' negative views of social media (Bennett et al., 2010). Social media can enable new behaviors that were previously difficult to achieve before these technologies reached the workplace, although corporate social media use research is still in its infancy (Leonardi & Treem, 2012). A study found that 86% of managers believe social media will be essential for their firm in three years, mostly for external communications (Kiron et al., 2012), yet not as much of work has investigated social media, especially enterprise social networks for internal communication with members of the organization or with others (Ellison et al., 2014). The usage of social media in the workplace can present both challenges and opportunities for employees, depending on how they choose to approach them (Oksa et al., 2022). Some researchers and organizations alike, contend that social media can interrupt work and subsequently affect employee performance. There is little scientific evidence to assist firms on permitting employees to use social media at work (Charoensukmongkol, 2014). Users risk adverse repercussions while using social media for professional purposes. One prominent feature is the widespread belief that using a personal media device intrudes on one's privacy and ultimately harms the user (Tarafdar et al., 2020). Also, work-related duties, and especially technology, particularly social media, increasingly follow employees home, where they can be a disruptive presence (Fieseler et al., 2015; Tams et al., 2020). However, compared to past technologies, enterprise adoption of social networking software is easy and prevention harder. This poses questions like: Are these sites productive? Do they increase enterprise utility? What new issues might new user populations face? How can these technologies be more productive given the changing landscape of work (remote work, AI and other technology at work)?

Future research should examine how much social and enterprise social networking sites are used for task-related versus social goals. Research must focus on interpreting if there are any variations in the use of these sites with regards to age, profession or function. With the increasing use of ESNS it would be necessary to study, design and theorize the ways in which using such tools at the workplace are helpful in transforming knowledge sharing and other organizational practices (Ellison et al., 2014). As a result, businesses that want to build internal knowledge networks can leverage a wide variety of social media to connect specialists and efficiently complement it with other forms of knowledge management software (Cao et al., 2016). Social media improves collaboration, which boosts Gen Y employees' learning and engagement (Naim & Lenka, 2017).

The theoretical model proposed through this review of existing literature provides insights on the different antecedents to Gen Y's use of social media. Several economic, cultural, technological, generational, organizational, as well as individual factors have been said to affect Gen Y's intention to use social media technology and its applications. Such uses of social media are known to ultimately fulfill work- related gratifications in the form of learning, knowledge sharing, employee engagement and bridging social capital at work. Thus, such avenues need to be provided to the employees by their organizations to allow a judicious use of these available resources.

CONCLUSION AND FUTURE RESEARCH

Due to the ever-increasing pace of change and fierce competition, it is essential to constantly update your knowledge and network. Because of the increasing number of millennials in the workforce, who value constant connectivity to the outside world, businesses may suffer if they do not offer this perk to their employees. As has been theorized by previous researchers, social media is used by employees to fulfill certain gratifications which may vary from individual to individual based on their individual, technological, or environmental factors. There are various problems that need to be answered by future research, despite the fact that the current study tries to determine the Uses and Gratifications of social media use:

- How much of the diversity may be explained by clustering over generations as opposed to simple aging?
- How can millennials be more productive?
- How can we use social media to attract millennials and retain their dedication and allegiance once we've hired them?
- What role may social media play in enhancing talent management in organizations?

Future research may explore these areas to complement the existing body of research concerning the U&G of social media at work.

From the standpoint of a company, it is crucial that management and HRD professionals understand the many advantages that SNS use may provide to employees and the firm as a whole. Improved work-related results for people and businesses as a whole may result from the implementation of social media policies, procedures, and systems.

REFERENCES

Andrés, L., & Cuberes, D. (2014). *The Diffusion of Internet : A Cross-Country Analysis*. Research Gate.

APAC: active social media penetration by country 2022. (n.d.). *Statista*. https://www.statista.com/statistics/255235/active-social-media-penetration-in-asian-countries/

Archambault, A., & Grudin, J. (2012). *A Longitudinal Study of Facebook, LinkedIn, & Twitter Use*. 2741–2750. ACM.

Ardichvili, A., Page, V., & Wentling, T. (2003). Motivation and barriers to participation in virtual knowledge-sharing communities of practice. *Journal of Knowledge Management*, *7*(1), 64–77. doi:10.1108/13673270310463626

Au, A. (2020). Reconceptualizing the generation in a digital(izing) modernity: Digital media, social networking sites, and the flattening of generations. *Journal for the Theory of Social Behaviour*, *50*(2), 163–183. doi:10.1111/jtsb.12239

Bagozzi, R. P., & Dholakia, U. M. (2002). Intentional social action in virtual communities. *Journal of Interactive Marketing*, *16*(2), 2–21. doi:10.1002/dir.10006

Bennett, K., Pitt, M., & Owers, M. (2010). Social Media Impact on Orginazations. In *5th Annual Conference Liverpool "BEAN."* Springer.

Kupperschmidt, B. (2000). *Multigeneration Employees: Strategies for Effective Management*. NIH.

Bolton, R. N., Parasuraman, A., Hoefnagels, A., Kabadayi, S., Gruber, T., Loureiro, Y. K., Migchels, N., & Solnet, D. (2013). *Journal of Service Management, 24*(3), 245–267.

Bolton, R. N., Parasuraman, A., Hoefnagels, A., Migchels, N., Kabadayi, S., Gruber, T., Loureiro, Y. K., & Solnet, D. (2013). Understanding Generation Y and their use of social media: A review and research agenda. *Journal of Service Management*, *24*(3), 245–267. doi:10.1108/09564231311326987

Boyd, D. & Ellison, N. B. (2008). Social Network Sites : Definition, History, and Scholarship. *Jcmc, 13*, 210–230. doi:10.1111/j.1083-6101.2007.00393.x

Calvo-Porral, C., & Pesqueira-Sanchez, R. (2020). Generational differences in technology behaviour: Comparing millennials and Generation X. *Kybernetes*, *49*(11), 2755–2772. doi:10.1108/K-09-2019-0598

Cao, X., Guo, X., Vogel, D., & Zhang, X. (2016). Exploring the influence of social media on employee work performance. *Internet Research*, *26*(2), 529–545. doi:10.1108/IntR-11-2014-0299

Castellacci, F., & Viñas-bardolet, C. (2019). *Internet use and job satisfaction*. Elesvier. doi:10.1016/j.chb.2018.09.001

Charoensukmongkol, P. (2014). Effects of support and job demands on social media use and work outcomes. *Computers in Human Behavior*, *36*, 340–349. doi:10.1016/j.chb.2014.03.061

Cheung, C. M. K., Chiu, P. Y., & Lee, M. K. O. (2011). Online social networks: Why do students use facebook? *Computers in Human Behavior*, *27*(4), 1337–1343. doi:10.1016/j.chb.2010.07.028

Cristea, G., Vatamanescu, E.-M., & Mitan, A. (2017). *Managing People Dynamics Through the Lens of Generation Y.* 11Th International Management Conference: The Role of Management in the Economic Paradigm of the Xxist Century (Imc 2017), *Bucharest, Romania.*

Cui, G., Lockee, B., & Meng, C. (2013). Building modern online social presence: A review of social presence theory and its instructional design implications for future trends. *Education and Information Technologies*, *18*(4), 661–685. doi:10.100710639-012-9192-1

Ellison, N. B., Gibbs, J. L., & Weber, M. S. (2014). The Use of Enterprise Social Network Sites for Knowledge Sharing in Distributed Organizations. *Sage*, *59*(1), 103–123. doi:10.1177/0002764214540510

Ellison, N. B., Steinfield, C., & Lampe, C. (2007). The Benefits of Facebook "" Friends :"' Social Capital and College Students '. *Journal of Computer-Mediated Communication*, *12*(4), 1143–1168. doi:10.1111/j.1083-6101.2007.00367.x

Fieseler, C., Meckel, M., & Ranzini, G. (2015). Professional Personae - How Organizational Identification Shapes Online Identity in the Workplace. *Journal of Computer-Mediated Communication*, *20*(2), 153–170. doi:10.1111/jcc4.12103

Gefen, D., & Straub, D. W. (2004). Consumer trust in B2C e-Commerce and the importance of social presence: Experiments in e-Products and e-Services. *Omega*, *32*(6), 407–424. doi:10.1016/j.omega.2004.01.006

Giannakos, M. N., Chorianopoulos, K., Giotopoulos, K., & Vlamos, P. (2013). *Behaviour & Information Technology Using Facebook out of habit. October*. Taylor & Francis. doi:10.1080/0144929X.2012.659218

Granovetter, M. (1983). The Strength of Weak Ties: A Network Theory Revisited. *Sociological Theory*, *1*, 201. doi:10.2307/202051

Haddud, A., Dugger, J. C., & Dugger, J. C. (2016). Social Media for Organizations. *Sage*, *3*(1).

Hargittai, E., & Hinnant, A. (2008). *Digital Inequality.*

Horng, S. M., & Wu, C. L. (2020). How behaviors on social network sites and online social capital influence social commerce intentions. *Information & Management*, *57*(2), 103176. doi:10.1016/j.im.2019.103176

Huub, R., Bondarouk, T., & Looise, K. J. (2023). E-HRM: Innovation or Irritation. An Explorative Empirical Study in Five Large Companies on Web-based HRM on JSTOR. *Management Review*, *15*(3), 364–380. https://www.jstor.org/stable/41783479

Immordino-yang, M. H., Christodoulou, J. A., & Singh, V. (2012). *Perspectives on Psychological Science Rest Is Not Idleness : Implications of.* Sage. doi:10.1177/1745691612447308

India: social media penetration 2022. (n.d.). Statista. https://www.statista.com/statistics/284436/india-social-media-penetration/

Khal, K., Bilgihan, A., Okumus, F., & Cobanoglu, C. (2013). Generation Y travelers ' commitment to online social network websites. *Tourism Management*, *35*, 13–22. doi:10.1016/j.tourman.2012.05.005

Kim, H. Lee, J., & Oh, S. E. (2019). Individual characteristics influencing the sharing of knowledge on social networking services: online identity, self-efficacy, and knowledge sharing intentions. *Sage*, *39*(4), 379–390. doi:10.1080/0144929X.2019.1598494

Kim, H., & Scott, C. (2019). Change communication and the use of anonymous social media at work: Implications for employee engagement. *Corporate Communications*, *24*(3), 410–424. doi:10.1108/CCIJ-07-2018-0076

Kim, W., Jeong, O. R., & Lee, S. W. (2010). On social Web sites. *Information Systems*, *35*(2), 215–236. doi:10.1016/j.is.2009.08.003

Kiron, D., Palmer, D., Phillips, A. N., & Kruschwitz, N. (2012). *Social Business: What Are Companies Really Doing?* Sage.

Kreutzer, T. (2009). *Generation Mobile : Online and Digital Media Usage on Mobile Phones among Low-Income Urban Youth in South Africa. February*. Research Gate.

Kwon, M.-W., D'Angelo, J., & McLeod, D. M. (2013). Facebook Use and Social Capital. *Sage*, *33*(1–2), 35–43. doi:10.1177/0270467613496767

Lampe, C., Ellison, N., & Steinfield, C. (2007). *A Familiar Face (book): Profile Elements as Signals in an Online Social Network*. 435–444.

Leonardi, P. M., & Treem, J. W. (2012). Knowledge management technology as a stage for strategic self-presentation: Implications for knowledge sharing in organizations. *Information and Organization*, *22*(1), 37–59. doi:10.1016/j.infoandorg.2011.10.003

Lowenthal, P. R. (2020). *Social Presence*. IGI Global. doi:10.4018/978-1-60566-984-7.ch011

Ma, W. W. K., & Chan, A. (2014). Knowledge sharing and social media: Altruism, perceived online attachment motivation, and perceived online relationship commitment. *Computers in Human Behavior*, *39*, 51–58. doi:10.1016/j.chb.2014.06.015

Macey, W. H. (2009). *Employee engagement : tools for analysis, practice, and competitive advantage*. Sage.

Magsamen-Conrad, K., Upadhyaya, S., Joa, C. Y., & Dowd, J. (2015). Bridging the divide: Using UTAUT to predict multigenerational tablet adoption practices. *Computers in Human Behavior*, *50*, 186–196. doi:10.1016/j.chb.2015.03.032 PMID:25937699

Mahmoud, A. B., Fuxman, L., Mohr, I., Reisel, W. D., & Grigoriou, N. (2021). "We aren't your reincarnation!" workplace motivation across X, Y and Z generations. *International Journal of Manpower*, *42*(1), 193–209. doi:10.1108/IJM-09-2019-0448

Massari, L. (2009). Analysis of MySpace user profiles. *Information Systems Frontiers 2009 12*(4), 361–367. doi:10.1007/s10796-009-9206-8

Meister, J. C., & Willyerd, K. (2010). *Mentoring Millennials: Spotlight on Leadership -- The Next Generation*. HBR. www.hbr.org

Melville, N., Kraemer, K., & Gurbaxani, V. (2004). Review: Information technology and organizational performance: An integrative model of it business value. *Management Information Systems Quarterly*, *28*(2), 283–322. doi:10.2307/25148636

Naim, M. F., & Lenka, U. (2017). Linking knowledge sharing, competency development, and affective commitment: Evidence from Indian Gen Y employees. *Journal of Knowledge Management*, *21*(4), 885–906. doi:10.1108/JKM-08-2016-0334

Ngai, E. W. T., Moon, K. L. K., Lam, S. S., Chin, E. S. K., & Tao, S. S. C. (2015). Social media models, technologies, and applications: An academic review and case study. *Industrial Management & Data Systems*, *115*(5), 769–802. doi:10.1108/IMDS-03-2015-0075

Nkwe, N., & Cohen, J. (2017). *Impact Of Social Network Sites On Psychological And Behavioural Outcomes In The Work-Place*.

Oksa, R., Pirkkalainen, H., Salo, M., Savela, N., & Oksanen, A. (2022). Professional social media-enabled productivity: A five-wave longitudinal study on the role of professional social media invasion, work engagement and work exhaustion. *Information Technology & People*, *35*(8), 349–368. doi:10.1108/ITP-11-2021-0899

Paroutis, S., & Al Saleh, A. (2009). Determinants of knowledge sharing using Web 2.0 technologies. *Journal of Knowledge Management*, *13*(4), 52–63. doi:10.1108/13673270910971824

Pribeanu, C., & Balog, A. (n.d.). *The 12 th International Scientific Conference eLearning and Software for Education Bucharest, Explaining The Acceptance Of Facebook By University Students With The Uses & Gratifications Theory*. eLSE. 10.12753/2066-026X-15-013

Quan-haase, A., & Young, A. L. (2010). *Bulletin of Science, Technology & Society Uses and Gratifications of of Facebook and Instant Messaging*. Sage. doi:10.1177/0270467610380009

Raacke, J., Ph, D., Bonds-raacke, J., & Ph, D. (2008). *MySpace and Facebook : Applying the Uses and Gratifications Theory to Exploring*. Mary Ann Liebert Pub. doi:10.1089/cpb.2007.0056

Rossmann, A., Ranjan, K. R., & Sugathan, P. (2016). Drivers of user engagement in eWoM communication Drivers of user engagement in eWoM communication (Journal of Services Marketing). *Journal of Services Marketing*, *C*(January), 541–553. doi:10.1108/JSM-01-2015-0013

Scheinbaum, A. C. (2017). A framework for the dark side of social media : From Digital Drama to Digital Over-engagement. *The Dark Side of Social Media*, 3–10. doi:10.4324/9781315167718-1

Shao, C., & Kwon, K. H. (2018). *Telematics and Informatics Clicks intended : An integrated model for nuanced social feedback system uses on Facebook. November*. Academia.

Shepherd, C. (2011). *Viewpoint Does social media have a place in workplace learning ? 27*(2), 3–4. doi:10.1108/02580541111103882

Sievert, H., & Scholz, C. (2017). Engaging employees in (at least partly) disengaged companies. Results of an interview survey within about 500 German corporations on the growing importance of digital engagement via internal social media. *Public Relations Review*, *43*(5), 894–903. doi:10.1016/j.pubrev.2017.06.001

Srinivasan, V. (2012). Multi generations in the workforce: Building collaboration. *IIMB Management Review*, *24*(1), 48–66. doi:10.1016/j.iimb.2012.01.004

Srite, M., & Karahanna, E. (2006). The role of espoused national cultural values in technology acceptance. *Management Information Systems Quarterly*, *30*(3), 679–704. doi:10.2307/25148745

Strohmeier, S. (2007). Research in e-HRM: Review and implications. *Human Resource Management Review*, *17*(1), 19–37. doi:10.1016/j.hrmr.2006.11.002

Sundar, S. S., & Limperos, A. M. (2013). *Uses and Grats 2.0: New Gratifications for New Media*. Taylor & Francis. doi:10.1080/08838151.2013.845827

Tams, S., Ahuja, M., Thatcher, J., & Grover, V. (2020). Worker stress in the age of mobile technology: The combined effects of perceived interruption overload and worker control. *The Journal of Strategic Information Systems*, *29*(1), 101595. doi:10.1016/j.jsis.2020.101595

Tarafdar, M., Pirkkalainen, H., Salo, M., & Makkonen, M. (2020). Taking on the "Dark Side" - Coping with Technostress. *IT Professional*, *22*(6), 82–89. doi:10.1109/MITP.2020.2977343

Taylor, P. E., & Keeter, S. E. (2010). *Millennials: Confident. Connected. Open to Change*. Pew Research Center.

Tiwari, S., Lane, M., & Alam, K. (2019). Do social networking sites build and maintain social capital online in rural communities? *Journal of Rural Studies*, *66*, 1–10. doi:10.1016/j.jrurstud.2019.01.029

van Puijenbroek, T., Poell, R. F., Kroon, B., & Timmerman, V. (2014). The effect of social media use on work-related learning. *Journal of Computer Assisted Learning*, *30*(2), 159–172. doi:10.1111/jcal.12037

Van Slyke, C., Bélanger, F., & Sridhar, V. (2005). A Comparison of American and Indian Consumers' Perceptions of Electronic Commerce. IGI Global. doi:10.4018/irmj.2005040102

Vyas, P., & Itam, U. J. (2022). *The Role of Social Media in Employee Knowledge Sharing*. IGI Global. doi:10.4018/978-1-6684-5849-5.ch009

Vyas, P. G., & Pandey, S. (2020). The effect of social networking sites use on employees' knowledge sharing. *European Journal of Training and Development*, *45*(2–3), 218–240. doi:10.1108/EJTD-02-2020-0027

Wang, S., & Wu, P. (2008). *Computers & Education The role of feedback and self-efficacy on web-based learning : The social cognitive perspective.* Science Direct. doi:10.1016/j.compedu.2008.03.004

Weyland, A. (2011). Engagement and talent management of Gen Y. *Industrial and Commercial Training*, *43*(7), 439–445. doi:10.1108/00197851111171863

Whiting, A., Williams, D., Whiting, A., & Williams, D. (2013). *Why people use social media : a uses and gratifications approach.* Emerald Insight. doi:10.1108/QMR-06-2013-0041

Yan, A., Wen, S., Vogel, D., & Kwok, R. C. (2010). *Computers & Education Can learning be virtually boosted ? An investigation of online social networking impacts.* Science Direct. doi:10.1016/j.compedu.2010.06.015

KEY TERMS AND DEFINITIONS

Bridging Social Capital: Bridging social capital refers to networks characterized by looser and weaker connections, typically involving individuals from diverse social networks with varying backgrounds and possessing distinct cultural, social, and economic resources. It encompasses the resources that an individual acquires through these networks.

Employee Engagement: Employee Engagement is the degree of a worker's enthusiasm and commitment towards their employment. It is the extent to which employees experience a mental and emotional connection to their work, their coworkers, and their organization.

Knowledge Sharing: Knowledge sharing is the exchange of knowledge (such as information, skills, and expertise) among individuals, acquaintances, colleagues, communities, or within or between organizations.

Social Media: Social- media refers to multimedia tools that enable individuals to create and disseminate knowledge, thoughts, passions, and various forms of communication within web-based networks and connections.

Work-related Learning: Work-related learning encompasses formal learning experiences that incorporate a designated period of instruction occurring within a professional environment or involving tasks directly associated with work contexts.

Chapter 6
Real-Time Bangla Sign Words Interpretation Uttered at Medical Emergencies

Tajbia Karim
Universiti Kebangsaan Malaysia, Malaysia

Zainal Rasyid Mahayuddin
Universiti Kebangsaan Malaysia, Malaysia

ABSTRACT

Sign language is a medium of communication using gestures which is developed for people who suffer from auditory or verbal impairment. But sign language is usually not widely familiar to common people. For this reason, a sign language interpreter is essential for translating sign word representation to regular text or voice. Though there are some research works on American Sign Language interpretation, but it is inadequate in number for other languages. This chapter illustrates real time interpretation of medical emergency relevant Bangla sign words using machine learning. Considering medical emergency situations, a real time word level interpreter is designed, and the performance is evaluated. The designed deep neural network uses Mediapipe Holistic network for feature detection from human body in transitional movement and Long short term memory (LSTM) to predict the image sequences of each sign action. The designed system provides 99% model training accuracy and 92.22% test accuracy in real time detection.

DOI: 10.4018/978-1-6684-8634-4.ch006

INTRODUCTION

Speech-impaired people are facing challenges every day to express their emotion and basic needs. In this world, around 5% people are suffering from speech impairment (Podder et al., 2022). Deaf and mute people are part of our society. It is not possible to achieve holistic sustainable development in social or economic sector leaving this huge population behind. Artificial intelligence can be applied to reduce inequality and attain well-being of speech impaired people. One of the important applications to achieve this goal is applying artificial intelligence for aiding deaf people in communication.

Sign language has evolved to facilitate these people to communicate with other people by means of hand gestures, facial expression and body language. Official formulation of Bangla sign language dataset is being developed by CDD (Center for Disability in Development) since the year 2000. There is only one organization titled Society of the Deaf and Sign Language Users (SDSL) which is educating people on Bangla sign language. Due to the lack of adequate facilities, the hearing impaired and verbally challenged people are still not being able to get full access in many vital sectors including education, healthcare and professional services (Siddiqua, 2014).

As common people are not familiar with sign language it is still a challenge for verbally challenged people to communicate effectively. At the time of emergency, the situation gets worse. In the past few decades researchers have come forward with the detection of Bangla sign language which mostly include Bangla sign alphabets. But this method is time consuming to express the proper sentence and thus is not appropriate for emergencies where time is constraint. The motivation of this research work is to facilitate verbally challenged people with an effective mean of communication in the time of emergencies.

This chapter discusses how artificial intelligence can be used to design a system that can interpret a Bangla sign word presented in front of camera in text format. As the system is designed specially for medical emergencies, hence a survey is performed to select nine most common words people may utter at a medical emergency situation. A training dataset is created where each sign word consists of sequential image frames. For fast pre-processing and key point extraction, Mediapipe Holistic is applied. Mediapipe Holistic is Google provided open-source framework that can be used for real-time applications. Sign words interpretation unlike sign alphabets, is not suitable to display only with one hand pose or gesture. It can require multiple steps or image frames to indicate one sign word. For this reason, sequence is very crucial to understand and interpret a sign word. In this research work, Long Short-Term Memory (LSTM) network is successfully applied to interpret each Bangla sign word from it's corresponding image sequence.

LITERATURE REVIEW

From one decade of research works, Wadhawan, A., & Kumar, P. mentioned in their review article that most of the research work (21%) related to sign language is performed on American Sign Language (Wadhawan & Kumar, 2021). In their article, Bangla sign language related research works are mentioned as very low in number and not even expressed as individual percentage value. Similarly, lack of research works on Bangla sign language detection is visible in the review article by Cheok et al. (2019) .

Sign language detection for medical emergency is essential for verbally challenged and hearing-impaired people. It allows them to express urgency using gestures. Also, people who are not familiar with sign language will face difficulties while communicating with sign language users. A system which can accurately detect and identify sign language words for medical emergencies is therefore of high importance. For Bangla language, sign detection at present is available for Bangla alphabets mostly. Several research have been published which used traditional image processing and computer vision methods. In recent times machine learning and deep learning-based methods have become a popular choice for sign language detection due to the higher accuracy and moderately easier implementation features. Convolutional Neural network was successfully used for feature detection in hand palm (Saif & Mahayuddin, 2023) .Yasir et al., (2018 implemented Convolutional Neural Network (CNN) for detecting Bangla alphabets. In another research Yasir et al., (2016) implemented SIFT based method along with SVM classifier to detect and identify hand gestures and train them for detecting words. But the research was done on regular Bangla sign words and using static images of hand gestures. Gestures with motion and from live feed was not taken as the input parameter for gesture detection. Saif and Mahayuddin, (2021) detected hand gesture using feature selection at important points of hand. Aminur Rahaman et al., n.d.) developed a modeling algorithm for Bangla sign language detection with different diversities including joint letters and dependent vowels. However, they have also used static images as gesture inputs. Uddin & Chowdhury, (2016) also used static hand gesture images to detect Bangla sign alphabets using Support Vector Machine (SVM). Islalm et al., (2019) used a CNN to train a dataset of 7052 images of Bangla sign alphabets. Basnin et al., (2021) utilized a CNN with LSTM layer to train a sign language recognition system using a dataset of 13,400 images of lexical signs. Haque et al., (2019) developed a system which can detect 26 gestures from Bangla sign language. They have used Principal Component Analysis (PCA) and K-nearest Neighbors to classify the gestures. Again, they have used static hand gesture images to train their model. Hasan et al., (2017) used SVM classifier to detect and identify gestures.

Shanta et al., (2018) used static hand gesture images for developing a system to detect and identify sign gestures. Deb et al., (2012) used computer vision-based method Normalized Cross Correlation two handed sign language detection. Islam et al., (2018) used CNN to train a model using the Eshara Lipi dataset to classify sign language numerals. Similar to previous researches they also used static gesture images. Alam et al., (2021) implemented a two dimensional CNN based approach to detect 36 Bangla alphabets and 10 numeric digits from hand gestures. Their system's input parameters were static hand gesture images. Angona et al. applied CNN based pre-trained network "MobileNet" for Bangla sign alphabet detection (Angona et al., 2020).

In recent research works, Katoch et al., (2022) detected Indian sign language in alphabet and digit form i.e., A- Z and 0-9. For feature extraction, they used SURF (Speeded Up Robust Features) and for classification, CNN and SVM were applied. Srivastava et al., (2022) demonstrated real-time Indian Sign Language detection using "TensorFlow Object Detection API". This model is also trained to detect alphabets and it is time consuming for recognizing an entire sign word. Another alphabet level recognition is presented by Rafi et al., (2019). Model uses CNN for classifying Bangla sign alphabets.

In the research article of Nihal et al., 2021, Bangla sign alphabets and 10 medical sign words are interpreted. For medical sign words, the selection process is not specified in the article. For feature detection, CNN is used and LSTM was trained to predict the sign word. The accuracy for medical word was achieved 87.5%. However, the sign words presentations were chosen from American Sign Language rather than Bangla sign language. Bangla sign word related research work lacks a lot for unavailability of sufficient dataset. Recently few researchers are coming forward by creating and publishing bangla sign word datasets. A recent inclusion is contributed by M. M. Islam et al. where the researchers presented 11 daily useful words such as "You", "Me", "Good" etc. (M. M. Islam et al., 2022).

Recognizing medical emergency words in real time poses several challenges. Lack of training dataset and proper motion tracking for gestures are two major challenges for detecting sign language words in real time. This research presents a novel approach for modelling of gesture motion and detecting medical emergency words in real time.

NOVELTY OF THIS RESEARCH

Previous researches that were done in the field of Bangla sign language detection involved both traditional computer vision based methods and machine learning based methods. The researchers developed several systems which can detect Bangla sign alphabets and Bangla sign words with high accuracy. However, to our knowledge, no research works were done to detect Bangla sign words for medical emergencies. Most Bangla sign language detection is designed for alphabet or digit level. A word level detection of 10 Bangla medical sign words were presented by Nihal et al., (2021) but the words were selected from American Sign Language (ASL). CNN was used for feature detection in this work. Our designed research method applied human gesture keypoint detection and tracking mechanism (with Mediapipe Holistic model) instead of CNN and achieved higher accuracy than the above-mentioned method.

This research takes motion-based hand gestures to detect sign words for medical emergencies in real time. This approach utilized a self-developed sequential gesture dataset to train a machine learning based model which involved gesture recognition to detect and identify the sign words for medical emergencies in real time. The developed model has demonstrated high accuracy in detecting real time test data from webcam feed.

METHODOLOGY

The research proceeded in following steps:

1. At first a survey has been conducted among 107 Bengali people to enlist the most uttered words in case of medical emergency.
2. From the survey result, the top nine words have been chosen for sign language interpretation.
3. With the kind support from **Centre for Disability in Development** (**CDD**), the Bangla sign language representation of these nine words have been collected in book and video clip format.
4. A dataset was created for chosen Bangla sign words. In this research work, 30 video clips are used per Bangla sign. Each of these video clips is comprised with 30 image frames. So, for each sign words, 900 image frames were collected and total database contain (900 X 9 = 8100) image frames .
5. Preprocessing and feature tracking is performed using Mediapipe holistic model. This model or network works faster than using CNN in all training images as it works in a principle that detects the keypoint features of human body (e.g., joints) in first image frame and then keeps tracking the same keypoints in consecutive images unless that body part is completely removed from the frame.
6. As sequence is very important in sign language representation, LSTM is applied to train the sequential data for each sign.
7. Then OpenCV has been used to predict Bangla signs in real time using webcam.

Survey Result

The motive of the survey was to find out the most probable words that a sick sign language user may have to use for communicating with general people or medical representatives. The survey questionnaire asked, "What are the most probable uttered words in case of medical emergencies such as sudden sickness at road or accidental case". The survey was conducted online in Google form format. The survey form was circulated among two university student and alumni groups of Bangladesh i.e., "AIUB Alumni Society" and ICT department students (05 batch) of Bangladesh University of Professionals (BUP). Voluntary response was requested for this research purpose. A total of 107 participants responded voluntarily in response to the form. As it was a public opinion survey, all these 107 responses are counted for selecting top nine medical emergency words. The age of the people was between 20-38 years and both female and male participants responded. The result of the survey is illustrated in Figure 1.

Bangla sign words relevant to medical emergency are initially collected from 'বাংলা ইশারা ভাষা অভিধান - Bengali Sign Language dictionary' published by National Centre for Special Education. The survey form contained some suggested words as well as an open comment section. The representation of top nine sign words selected after the survey. The representation is illustrated in Figure 2. Arrow represents the direction of hand movement in each sign word.

The Methodology is illustrated with Figure 3. Figure 3 describes the entire process in brief. Within the process, the working method of feature detection by Mediapipe is illustrated as well.

In this research, LSTM network is trained for sequential data prediction. LSTM network contains different units. Figure 4 illustrates the working method of LSTM network.

Figure 1. Survey result revealing most uttered words at medical emergency.

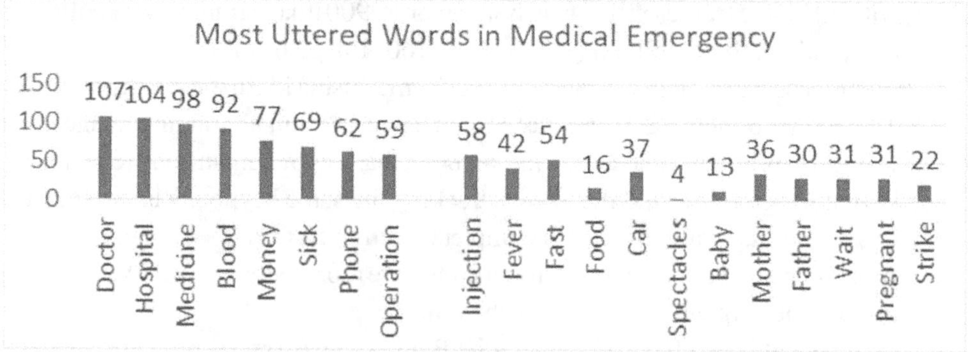

Figure 2. Pictorial description of the most uttered Bangla Sign words at medical emergencies
Image source: 'বাংলা ইশারা ভাষা অভিধান - Bengali Sign Language dictionary' published by National Centre for Special Education.

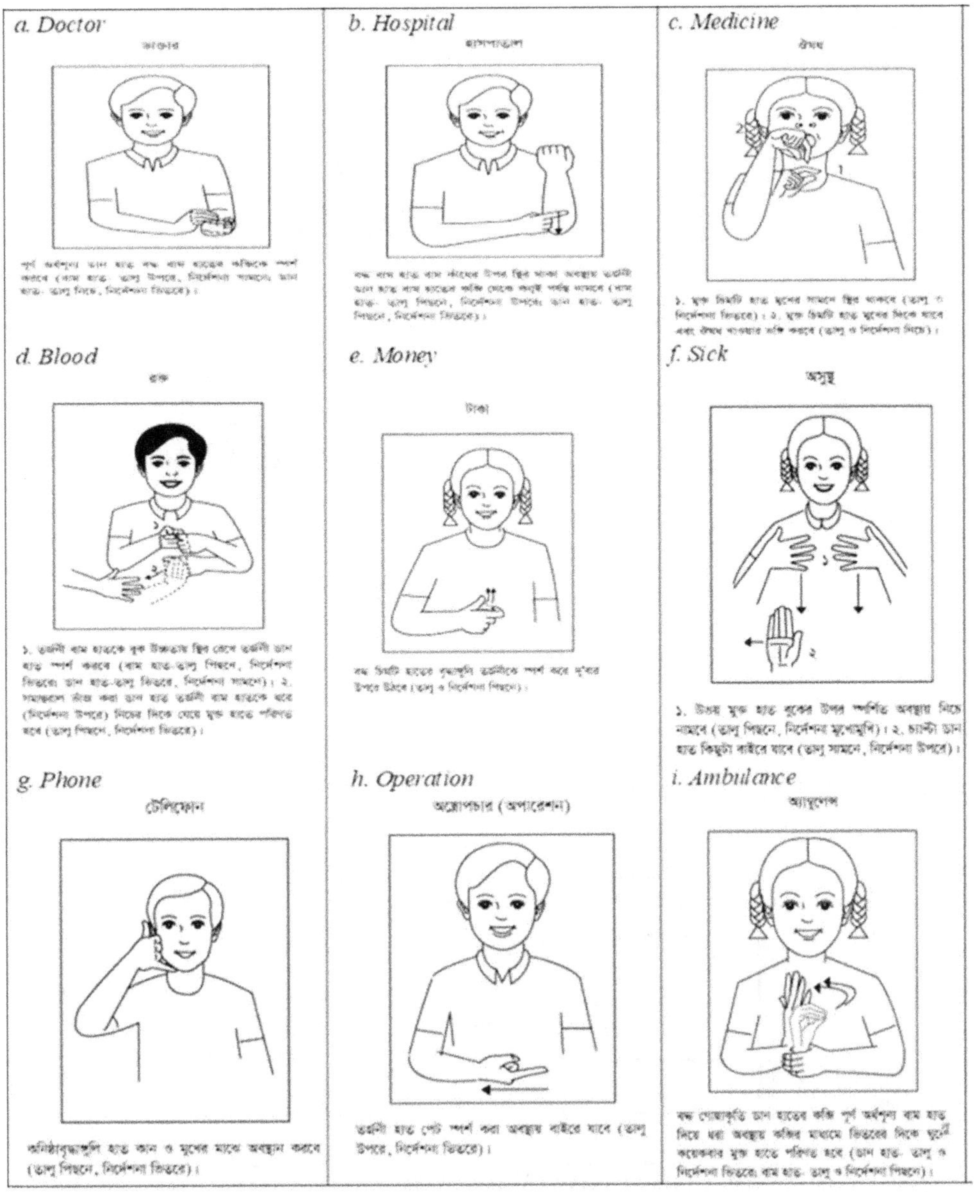

Figure 3. Block diagram of the methodology

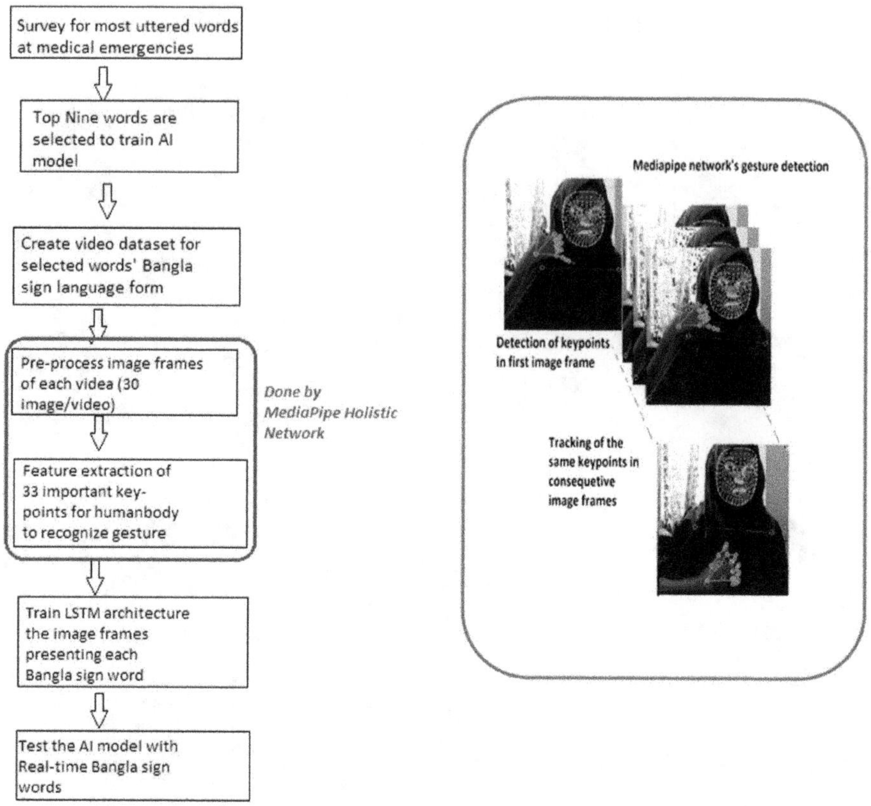

Figure 4. Working method of LSTM network

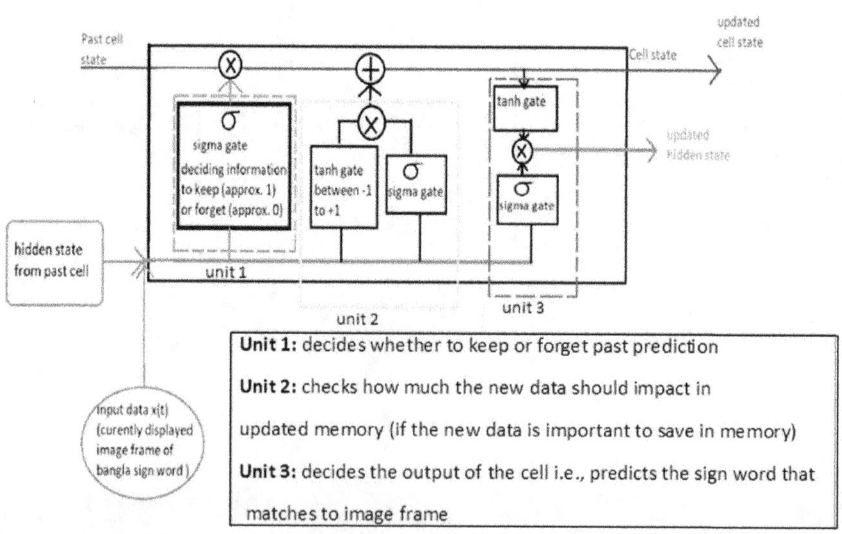

Mediapipe holistic is applied instead of CNN network as it does not detect features in every image frame. It simplifies the process by detecting key features in the first image frame then tracking those exact features in consecutive image frames making the learning parameter lower. This is described in Figure 3.

Since each sign word is presented with some sequential movement of hand, a sequential detection network is essential in this research. Thus LSTM is added to the network. In Figure 4, the detailed process of LSTM is described in pictorial method. LSTM is comprised of several gates. The orange unit in the picture is called forget gate. If an image frame from previous time stamp is no longer matching with the context of new image frame, (such as step for 'Doctor' was being shown. Suddenly the presenter is showing 'money'), then this forgets the previous detection. The yellow unit uses a tanh gate that decides how much the new input should be kept in the LSTM cell, i.e., how essential it is. The purple unit chooses what prediction to output based on the sequences of input. The cell state hence carries a long term memory by updating with the essential predictions based on sequential context.

EXPERIMENTAL SETUP

The research was conducted with Intel(R) Core(TM) i7-7500U CPU 64-bit operating. The programming language is python. For this research, Mediapipe Holistic is used to detect Sign Language signs. MediaPipe Holistic is used to extract key points from hand, body and face. The key point locations are described in Figure 5. Mediapipe Holistic model is comprised of 3 models: Hand, pose and face.

Afterwards, TensorFlow in Keras is used to build a LSTM model to predict the action which is shown in screen. These actions present the sign language sign words. At the end the trained model is used for Bangla signs are prediction in real time using a webcam. The detection output is in text format.

Model Architecture

The model description is illustrated in Figure 6. Here softmax is used as activation function and the model is optimized with Adam optimizer to get optimized learning rate. Each time is the model is trained using 3 sets of words such as first batch: Doctor, hospital medicine; second batch: Blood, money, sick ; third batch: Phone, ambulance, operation. The reason behind it is, the model needs 900 image frames per sign word. With all the 9 words trained (81000 images) the processor had to handle a tremendous amount of training parameters. To conveniently perform the experiment with lower number of parameters, each time 3 sign words (900 X3 = 2700 image frames) were trained to the designed model and tested.

Figure 5. Mediapipe Holistic keypoint locations (a) Hand landmark model (Hand Landmarks Detection Guide | MediaPipe | Google Developers, n.d.)(b) Pose Landmark Model (On-Device, Real-Time Body Pose Tracking with MediaPipe BlazePose – Google AI Blog, n.d.) (c) Face Landmarks(Face Landmark Detection Guide | MediaPipe | Google Developers, n.d.)

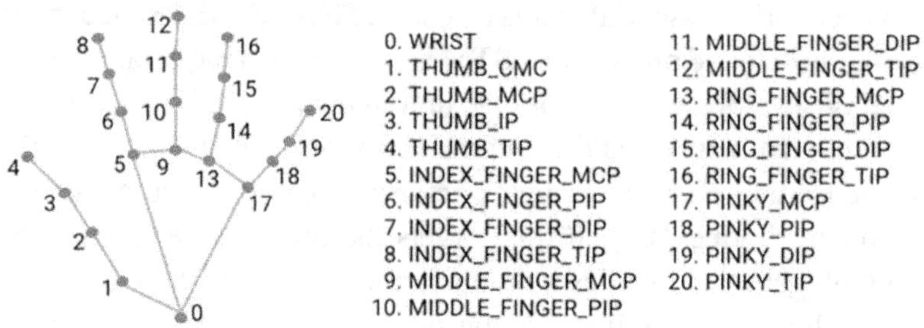

(a) (*Hand Landmarks Detection Guide | MediaPipe | Google Developers*, n.d.)

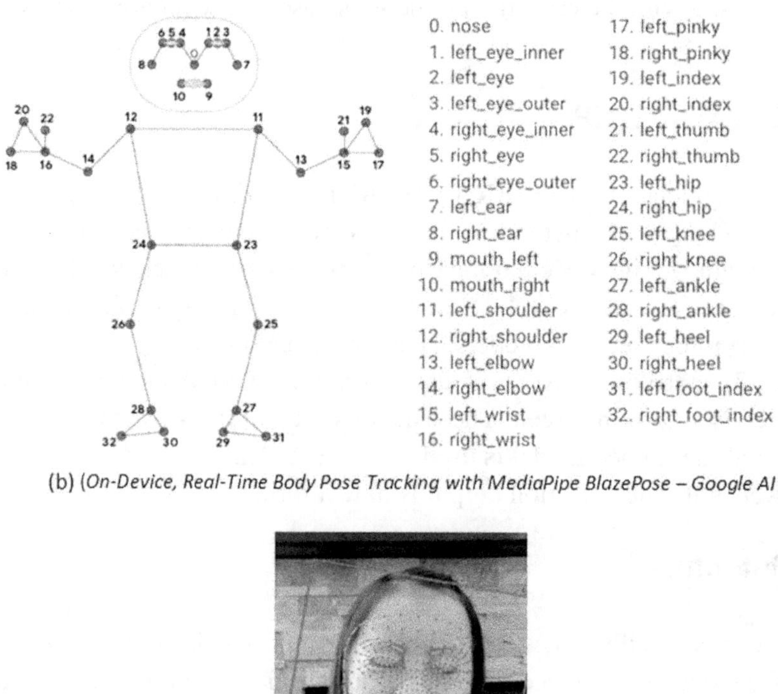

(b) (*On-Device, Real-Time Body Pose Tracking with MediaPipe BlazePose – Google AI Blog*, n.d.)

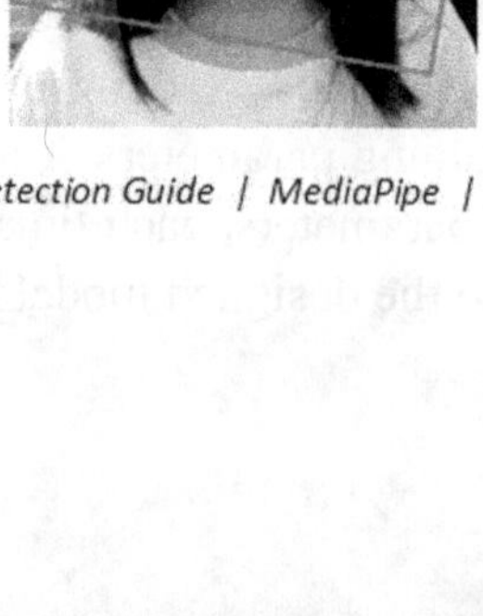

(c) (*Face Landmark Detection Guide | MediaPipe | Google Developers*, n.d.)

Figure 6. Model architecture

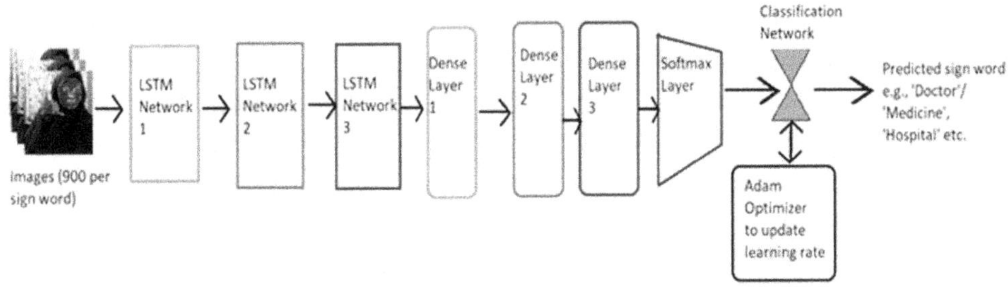

RESULT WITH REAL-TIME TEST DATA PREVIOUSLY UNSEEN TO THE MODEL

After training the model, it is tested with continuous real time data using webcam. The results are illustrated as follows. The detection latency is extremely low. Different user and variable light condition is used for testing. Based on the presented steps in front of camera, the detection or predicted sign word is highlighted. Figure 7 presents the real-time test data of different Bangla sign words. From the demonstrated steps, the model predicted that user wants to say "doctor" in Figure 7(a). Based on the sequence of the steps the sign word is predicted by trained LSTM network. Similarly, for sequential steps of Figure 7(b), "Hospital" is predicted and so on.

For sequential detection, in this research, LSTM network is used. This LSTM network predicts based on the sequences of steps it gets at input (via webcam in

Figure 7. Real time detection of Bangla signs (a) Doctor (b) Hospital (c) Medicine (d) Blood (e) Money (f) Sick (g) Phone (h) Operation (i) Ambulance using Webcam

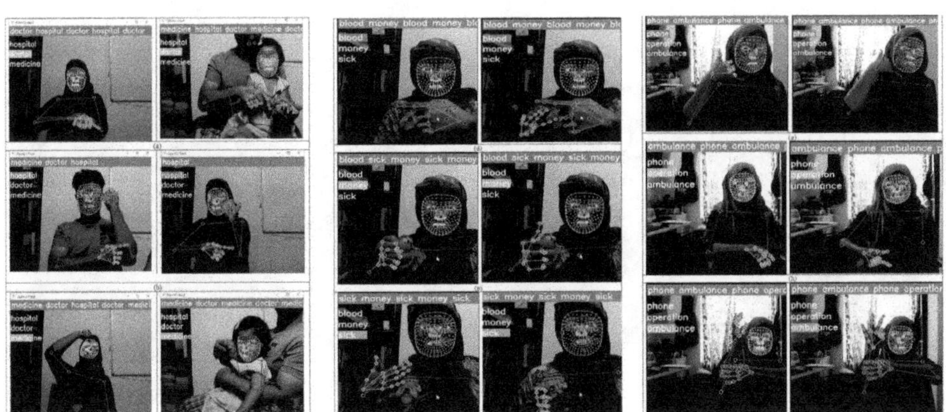

this case) and generates predicted sign word class accordingly. It appears like real-time due to the very low latency. The following link shares a sample test session: https://drive.google.com/file/d/1nO7s0ZCV7uHd-GvD5fYYMFM9p7NVhTdN/view?usp=drive_link

For training purposes, 126 epoch is found sufficient. The training accuracy reaches 100 and loss to 0 at about 126 epochs. An interesting fact is noticed while training the data clips i.e., training accuracy almost saturates after 126 epochs. Moreover, excess training with the same amount of data reduces the accuracy. Figure 8 represents this phenomenon perfectly. It is clearly visible that after 1800 epochs, accuracy decreases and loss increases considering the number of training data remain same. This phenomenon occurs due to 'Overfitting' (Brownlee, 2021). This issue is also depicted from the accuracy and loss curve presented in Figure 8.

After completion of training the LSTM model, it is tested for continuous real time webcam video data. Each sign word was displayed 20 random moment and the prediction was noted. The accuracy for real time test data is found to be 92.22%. The details are illustrated in shown in Table 1. The participants were in different lighting conditions.

Mostly the incorrect detection occurs at the beginning of demonstrating a new sign word. For example after demonstrating 'blood', if 'money' sign word is displayed without giving any time delay, this will be detected incorrectly. Since, this model uses sequential prediction model, it would need some time when a new sign word is displayed in front of the webcam to match with the consecutive sequence and predict by LSTM.

It is found that sign words which mostly focus of detailed hand tracking than holistic pose, such as the word "Money", provide lower accuracy while testing in real-time. The reason behind is, pose can be estimated with lower resolution

Figure 8. Model accuracy with respect to epochs and training loss with respect to epochs

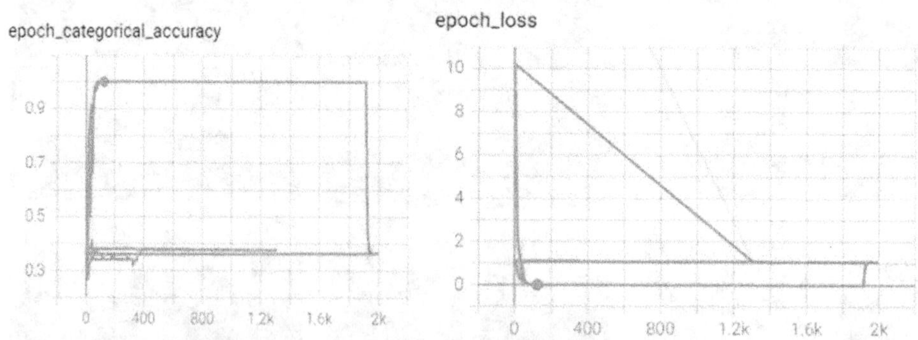

Table 1. Real-time accuracy result in test session

Class	Correct Detection	Incorrect Detection	Accuracy
Doctor	18	2 (Hospital)	90%
Hospital	20	-	100%
Medicine	19	1(doctor)	95%
Blood	19	1 (money)	95%
Money	15	3(Blood) 2(Sick)	75%
Sick	18	2 (Money)	90%
Phone	20	-	100%
Operation	18	2(phone)	90%
Ambulance	19	1(operation)	95%
Total: 180	166	14	92.22%

(256X256 pixel) video frame. But when it requires to crop hand region to pass to it's respective model, it appears to be inadequate for accurate detection.

A comparison of our implemented method with the relevant recent methods is shown Table 2.

LIMITATION AND FUTURE WORK

Predicting sequential data is a complex task. As each sign word prediction is dependent on sequences of hand / body gestures, it is essential to assign a particular number

Table 2. Comparison of the implemented method with relevant recent methods

Methodology	Detection	accuracy	Suitability to use at medical emergency
(Katoch et al., 2022)	Alphabet and digit	99%	No (since it is time consuming)
(Srivastava et al., 2022)	Alphabet	93-96%	Time consuming
(Nihal et al., 2021)	Alphabet And word	87.5% for medical sign words	Yes
(Rafi et al., 2019)	Alphabets	89.6%	No (since it is time consuming)
Proposed method	Sign words related to medical emergency	training accuracy 99% real time test accuracy 92.22%	Yes

of image frames for classifying each category while training the LSTM model. In this research 30 image frames per video is applied for each training sample of sign word. As LSTM matches the consecutive sequences of image frames for generating a prediction, for sign words with very few steps, such as "Money" sign word achieved lower accuracy compared to other sign words (comparison is shown in tabe 1). Specially, at first try user was not successful for getting correct prediction. It needed repetition for generating correct prediction.

One solution to achieve better accuracy in such words can be by enhancing dataset. In future work, the researchers intend to include more medical emergency sign words and enhance the dataset size.

CONCLUSION

This research has been inspired to minimize the barrier of communication between a speech impaired person and general people with the help of artificial intelligence. The focus of this research is Bangla sign words interpretation in text format. The entire work is concentrated on medical emergency relevant words such as sudden sickness or road accident situation where a sign language user wants to communicate to common people or medical representatives.

The presented medical emergency relevant words are chosen based on a survey among 107 Bangladeshi people. The biggest challenge was to train the model with available dataset since the available data are in image format representing Bangla sign alphabets e.g., the most used dataset Ishara-Lipi (Sanzidul Islam et al., 2018). But we needed a dataset where each training data is in video clip format expressing the Bangla sign words accurately. With this purpose a training database was created with the kind support from the video clip instructions from CDD. The dataset we have built contains 270 video clips each having 30 image frames (in total 8100 image frames).

While conducting this research it is realized that Bangla sign language related research works are very few in number. The few works conducted in this sector, mostly are focused on sign alphabet interpretation. For medical emergency situation, sign word interpretation is more necessary for the time constraint. This research work has interpreted nine most pronounced words in case of medical emergency with an average of 92.22% test accuracy even with low lit condition. With the expansion of dataset, the model can be used to identify more Bangla sign language signs and contribute to the wellbeing of speech impaired people of Bangladesh.

ACKNOWLEDGMENT

This research work is financially supported by following two research grants:

i. Research grant of Center for Higher Studies, Bangladesh University of Professionals, supported by University Grants Commission, Bangladesh.
ii. The "Geran Universiti Penyelidikan" research grant, GUP-2020-064 by Universiti Kebangsaan Malaysia

The authors would also like to acknowledge the kind support from Iftekhar Ahmed Shorful, Coordinator of The Centre for Disability in Development (CDD) who has provided the rare pictures and videos of Bangla sign words to create the training dataset.

REFERENCES

Alam, M. S., Tanvir, M., Saha, D. K., & Das, S. K. (2021). Two Dimensional Convolutional Neural Network Approach for Real-Time Bangla Sign Language Characters Recognition and Translation. *SN Computer Science*, *2*(5), 1–13. doi:10.100742979-021-00783-6

Aminur, R. A. (2020). M., Jasim, M., Haider ALI, M., & Hasanuzzaman, M. (n.d.). Bangla language modeling algorithm for automatic recognition of hand-sign-spelled Bangla sign language. *Frontiers of Computer Science*, *143302*(3), 143302. doi:10.100711704-018-7253-3

Angona, T. M., Shaon, A. S. M. S., Rashad Niloy, K. T., Karim, T., Tasnim, Z., Reza, S. M. S., & Mahbub, T. N. (2020). Automated bangla sign language translation system for alphabets by means of MobileNet. [Telecommunication Computing Electronics and Control]. *Telkomnika*, *18*(3), 1292–1301. doi:10.12928/telkomnika.v18i3.15311

Basnin, N., Nahar, L., & Hossain, M. S. (2021). An integrated cnn-lstm model for bangla lexical sign language recognition. *Advances in Intelligent Systems and Computing*, *1309*, 695–707. doi:10.1007/978-981-33-4673-4_57

Brownlee, J. (2021). *Machine Learning Mastery With Python: Understand Your Data, Create Accurate.* Google Books. https://books.google.com.my/books?hl=en&lr=&id=BgmqDwAAQBAJ&oi=fnd&pg=PP1&dq=J+Brownlee+-+Machine+Learning+Mastery,+2016&ots=frp-TtkJ_O&sig=RLdyBCyOZ9EHKol45PLPV0uO0Ig&redir_esc=y#v=onepage&q&f=false

Cheok, M. J., Omar, Z., & Jaward, M. H. (2019). A review of hand gesture and sign language recognition techniques. *International Journal of Machine Learning and Cybernetics*, *10*(1), 131–153. doi:10.100713042-017-0705-5

Deb, K., Mony, H. P., & Chowdhury, S. (2012). Two-Handed Sign Language Recognition for Bangla Character Using Normalized Cross Correlation. *Global Journals.*

Face landmark detection guide. (n.d.). Google Developers. https://developers.google.com/mediapipe/solutions/vision/face_landmarker/

Haque, P., Das, B., & Kaspy, N. N. (2019, April 1). Two-Handed Bangla Sign Language Recognition Using Principal Component Analysis (PCA) and KNN Algorithm. *2nd International Conference on Electrical, Computer and Communication Engineering, ECCE 2019*. IEEE. 10.1109/ECACE.2019.8679185

Hasan, M., Sajib, T. H., & Dey, M. (2017, January 26). A machine learning based approach for the detection and recognition of Bangla sign language. *1st International Conference on Medical Engineering, Health Informatics and Technology, MediTec 2016*. IEEE. 10.1109/MEDITEC.2016.7835387

Islalm, M. S., Rahman, M. M., Rahman, M. H., Arifuzzaman, M., Sassi, R., & Aktaruzzaman, M. (2019, September 1). Recognition Bangla Sign Language using Convolutional Neural Network. *2019 International Conference on Innovation and Intelligence for Informatics, Computing, and Technologies (3ICT)*. IEEE. 10.1109/3ICT.2019.8910301

Islam, M. M., Uddin, M. R., Ferdous, M. J., Akter, S., & Nasim Akhtar, M. (2022). BdSLW-11: Dataset of Bangladeshi sign language words for recognizing 11 daily useful BdSL words. *Data in Brief*, *45*, 108747. doi:10.1016/j.dib.2022.108747 PMID:36425983

Islam, M. S., Mousumi, S. S. S., Azad Rabby, A. K. M. S., Hossain, S. A., & Abujar, S. (2018). A Potent Model to Recognize Bangla Sign Language Digits Using Convolutional Neural Network. *Procedia Computer Science*, *143*, 611–618. doi:10.1016/j.procs.2018.10.438

Katoch, S., Singh, V., & Tiwary, U. S. (2022). Indian Sign Language recognition system using SURF with SVM and CNN. *Array (New York, N.Y.)*, *14*, 100141. doi:10.1016/j.array.2022.100141

Nihal, R. A., Broti, N. M., Deowan, S. A., & Rahman, S. (2021). Design and Development of a Humanoid Robot for Sign Language Interpretation. *SN Computer Science*, *2*(3), 1–17. doi:10.100742979-021-00627-3

On-device, Real-time Body Pose Tracking with MediaPipe BlazePose – Google AI Blog. (n.d.). Retrieved May 15, 2023, from https://ai.googleblog.com/2020/08/on-device-real-time-body-pose-tracking.html

Podder, K. K., Chowdhury, M. E. H., Tahir, A. M., Mahbub, Z., Khandakar, A., Hossain, M. S., & Kadir, M. A. (2022). Bangla Sign Language (BdSL) Alphabets and Numerals Classification Using a Deep Learning Model. *Sensors (Basel)*, *22*(2), 574. Advance online publication. doi:10.339022020574 PMID:35062533

Rafi, A. M., Nawal, N., Bayev, N. S. N., Nima, L., Shahnaz, C., & Fattah, S. A. (2019, October 1). Image-based Bengali Sign Language Alphabet Recognition for Deaf and Dumb Community. *2019 IEEE Global Humanitarian Technology Conference, GHTC 2019*. IEEE. 10.1109/GHTC46095.2019.9033031

Saif, A., & Mahayuddin, Z. R. (2021). An efficient method for hand gesture recognition using robust features vector Authors AFM Saif, Zainal Rasyid Mahayuddin. *Journal Information System and Technology Management (JISTM)*. https://scholar.google.com/citations?view_op=view_citation&hl=en&user=1Y50UjsAAAAJ&sortby=pubdate&citation_for_view=1Y50UjsAAAAJ:hFOr9nPyWt4C

Saif, A., & Mahayuddin, Z. R. (2023). View artiAugmented Reality-Based 3D Human Hands Tracking From Monocular True Images Using Convolutional Neural Networkcle. *Handbook of Research on Artificial Intelligence and Knowledge Management in Asia's Digital Economy*. IGI Global. https://scholar.google.com/citations?view_op=view_citation&hl=en&user=1Y50UjsAAAAJ&sortby=pubdate&citation_for_view=1Y50UjsAAAAJ:e5wmG9Sq2KIC

Sanzidul Islam, M., Sultana Sharmin Mousumi, S., Jessan, N. A., Shahariar Azad Rabby, A., & Akhter Hossain, S. (2018, November 30). Ishara-Lipi: The First Complete MultipurposeOpen Access Dataset of Isolated Characters for Bangla Sign Language. *2018 International Conference on Bangla Speech and Language Processing, ICBSLP 2018*. IEEE. 10.1109/ICBSLP.2018.8554466

Shanta, S. S., Anwar, S. T., & Kabir, M. R. (2018, October 16). Bangla Sign Language Detection Using SIFT and CNN. *2018 9th International Conference on Computing, Communication and Networking Technologies, ICCCNT 2018*. IEEE. 10.1109/ICCCNT.2018.8493915

Siddiqua, F. (2014, August 22). The Silent Conversation. *The Daily Star*. https://www.thedailystar.net/the-silent-conversation-37929

Srivastava, S., Gangwar, A., Mishra, R., & Singh, S. (2022). Sign Language Recognition System using TensorFlow Object Detection API. *Communications in Computer and Information Science, 1534 CCIS*, 634–646. doi:10.1007/978-3-030-96040-7_48

Uddin, A., & Chowdhury, S. A. (2016, February 14). Hand sign language recognition for Bangla alphabet using Support Vector Machine. *2016 International Conference on Innovations in Science, Engineering and Technology (ICISET)*. IEEE. 10.1109/ICISET.2016.7856479

Wadhawan, A., & Kumar, P. (2021). Sign Language Recognition Systems: A Decade Systematic Literature Review. *Archives of Computational Methods in Engineering*, *28*(3), 785–813. doi:10.100711831-019-09384-2

Yasir, F., Prasad, P. W. C., Alsadoon, A., & Elchouemi, A. (2016). SIFT based approach on Bangla sign language recognition. *2015 IEEE 8th International Workshop on Computational Intelligence and Applications, IWCIA 2015 - Proceedings*, (pp. 35–39). IEEE. 10.1109/IWCIA.2015.7449458

Yasir, F., Prasad, P. W. C., Alsadoon, A., Elchouemi, A., & Sreedharan, S. (2018). Bangla Sign Language recognition using convolutional neural network. *2017 International Conference on Intelligent Computing, Instrumentation and Control Technologies, ICICICT 2017, 2018-January*, (pp. 49–53). IEEE. 10.1109/ICICICT1.2017.8342533

KEY TERMS AND DEFINITIONS

Sign Word: Words that are conveyed with hand or body gesture or expression. Especially applied by verbally challenged people

MediaPipe Holistic: A framework developed by Google that is made opensource for realtime detection of body gesture

LSTM: This stands for long short term memory network, i.e., applied for sequential data prediction.

CNN: This stands for convolutional neural network, which is used for pattern prediction.

OpenCV: A computer vision library.

Chapter 7
An Empirical Study on the Impact of Social Media Promotion of Branded Drugs on the Health Behavior of Indian Patients

Samrat Kumar Mukherjee
https://orcid.org/0000-0002-8839-0140
Sikkim Manipal University, India

Abhijit Sarkar
https://orcid.org/0000-0002-1042-2712
SRM University, India

Subhadeep Mukherjee
https://orcid.org/0000-0003-2138-9089
Ramaiah University of Applied Sciences, India

Jitendra Kumar
https://orcid.org/0000-0003-0167-0738
Sikkim Manipal Institute of Technology, India

Ajeya Jha
https://orcid.org/0000-0003-0491-5008
Sikkim Manipal Institute of Technology, India

ABSTRACT

Branded drug social media promotion (SMP) has increased dramatically in the recent era. The impact of this new phenomenon (although it is illegal in most countries) on patient behaviour has not been thoroughly studied. The purpose of this research is to learn about patients' perceptions of the impact of SMP on their behaviour. The sample size is 380 people. Pearson correlation and regression were used for analysis. The study clearly shows that SMP has an effect on patients' negative behaviour as well as the patient-physician relationship. The government and physicians should be conscious that proper guidelines are needed to guide patients. In addition, patients must understand that physicians and their expertise are critical to their treatment.

DOI: 10.4018/978-1-6684-8634-4.ch007

INTRODUCTION

Usage of social media platforms for health information has increased its importance in recent past. Disastrous effect of covid-19 pandemic has also influenced common people to increase their traffic in various social media platforms like - Facebook, twitter, YouTube. Changing lifestyles followed by increasing stress and a try to stay connected with the virtual world has forced people to stay glued into social media platforms. It has been observed in many research that at present normal public, patients, health professional and health service provider Statistica report has observed a gradual increasing trend in social media penetration. In India people using internet and access to social media platforms has increased from 0.97 billion to 2.65 billion from 2010 to 2018 and it is projected to increase to over 3 billion in 2018 (Report from Statistia, 2020) Fetching of health-related information from the social media platforms has been the recent trend. It has gained importance and popularity because of ease of accessing multiple health related information on variety of health-related topics. Social media health information has not only helped the netizens to access multiple health related information at a click but also it has been helped them to stay emotionally connected to other people having the same medial problems or similar health issues (Zhao & Zhang, 2017). Popular social media platforms like Instagram, Facebook, Twitter, YouTube, and other social media platform provide patients an opportunity to become an active and informed customer (Benetoli., 2017). It acts as a ready reckoner of health related videos, blogs for the patients and public. Patients may not always look for information, but also contribute their expertise, experiences, and symptoms about health care. Social media platforms may also be used to post reviews on health items, drugs, and physicians. Patients with comparable ailments can build bonds and create health communities and groups Social media connections provide patients with information, emotional support, and social interaction, allowing them to manage with their disease. By the second week of the lockdown (March 28 to April 3), time spent on social networking apps had grown to 5 hours per week in India (as per a recent statistica report, 2020). Positive messages on various networks helped individuals stay upbeat and encouraged. However, there have been allegations that medical misinformation and unconfirmed content are also creating uncertainty and alarm. As a result, many social media platforms such as Facebook and Twitter are directing users to trustworthy sources such as WHO websites and local health authority websites.

Pharmaceutical products in the market are broadly categorized as branded drugs and generic drugs. Branded drugs, also referred to as brand-name or innovator drugs, are developed, patented, and marketed by a specific pharmaceutical company. These drugs are protected by patents that grant exclusive rights to the manufacturer for a certain period. Branded drugs are introduced with a unique brand name and are often associated with significant investments in research and development. On the other hand, generic drugs are developed and brought to the market after the expiration of patents protecting the original branded drugs. These drugs contain the same active ingredients as their branded counterparts, have identical dosage forms, strengths, and routes of administration. However, generic drugs are generally marketed using their non-proprietary name instead of a brand name (Food and Drug Administration, 2020).

In the digital era, social media has become a significant platform for communication, sharing information, and facilitating engagement. The pharmaceutical industry has acknowledged the potential of social media as a means to promote branded drugs and reach a broader audience. Utilizing social media for promotion presents pharmaceutical companies with unique opportunities to directly connect with consumers, healthcare professionals, and patient communities, enabling the provision of valuable information and fostering engagement (Mackey et al., 2015). However, alongside the numerous advantages of social media promotion, there are challenges that need to be addressed. These challenges encompass adhering to regulatory requirements, ensuring the dissemination of accurate and unbiased information, managing potential adverse events, and effectively handling user-generated content. By successfully navigating these challenges, pharmaceutical companies can leverage the power of social media to establish brand trust, enhance patient engagement, and drive positive health outcomes (Smith et al., 2022).

LITERATURE REVIEW

According to IAMAI and IMRB (2013), the time spent by Indian netizens on social media platforms varies depending on the day of the week. On weekdays, the average time spent is 29.6 minutes, while on weekends, it slightly decreases to 28.8 minutes. The majority of social media users in India are concentrated in metropolitan areas. Platforms like Facebook, Twitter, LinkedIn, and YouTube are commonly used for promotional purposes. In 2022, the Indian digital advertising market is projected to become the largest medium, accounting for 45% of the market share (Economic Times, 2022). The number of internet users in India reached 749 million in 2020, and it is estimated to surpass 1.9 billion by 2040 (Statista, 2022). This growing internet user base has fueled the expansion of various digital businesses, including e-commerce and digital advertising. India holds the second-largest internet community in the world, following China.

Concern for own health care is forcing patients to become more interested in medication information. Some of these information are institutional and in character, with the goal of improving the company's goodwill or professional image rather than promoting one or more specific items. Other advertisements were more explicit about ailment or medication treatment and medicine. This sort of promotional strategy is intended to promote physician-patient engagement and, as a result, demand for branded medicines (Cutrer & Pleil, 1991).

Rising demand for medication and health information through social media is forcing few people with underhand intention. (Kyytsonen et al., 2022; Papapgeorgiou et al., 2018) have observed in their research after an analysis of mobile health apps, that a huge number of applications are unable to secure users' sensitive health information from third-party use, processing, or access. Utilizing social network for health information, can lead to more equitable discussion among patients. Resulting in increased trust on the healthcare practitioner. Patients can learn more about therapy choices by using information from social media networks. (Wicks et al., 2010) documented in their study that individuals are better equipped with informed knowledge to interact with healthcare professionals since they have a better understanding of their situation, resulting in patients to feel more confident in their doctor-patient interaction (van et al., 2008, Barlett & Coulson, 2011). Barlett and Coulson (2011) argued in their paper that due to increased access to medical knowledge patients feel more prepared for consultations because they are more informed about their disease and know what questions to be asked; they also feel more powerful within. Media exposure also increases the chance of having a desire to actively communicate with a clinician during a medical consultation (Oh & Lee, 2012). Furthermore, social media usage allows a patient to develop and enhance their communication skills with respect to their health, resulting increased patient's demand for medical care treatment (Lee & Wu, 2014). Francisco Jose Grajales (2014) classified social media platforms into 10 groups in his research. For each category, the survey supplied the names of prominent social networking sites. When looking for particular platform names, it is vital to remember that all 13 research covered in the table listed Facebook as a social media platform utilised for health care communications. Patients were able to follow health-related pages on Facebook. Patients can also take part in disease-specific conversations. Social media has now changed the approach of looking after a health or mental health issues by a patient. It has evolved into a standard online platform that allows people to connect, produce, and govern material across time and location among multiple users. Patients mostly use social media to seek information about a condition and its treatment (Zhao et al., 2017, Benetoli et al., 2017, Bhaskaran et al., 2017), as well as to seek the knowledge and experiences of other patients (Antheunis et al., 2013, Al-Qahtani et al., 2018, Greene et al., 2011) to seek emotional support (Zhao et al., 2017, Benetoli et al., 2017, Smailhodzic et al., 2016), to share personal health stories and experiences (Moorhead et al., 2013, Lefebvre et al., 2013, Oh HJ et al., 2013), to seek and post reviews of medical professionals and service providers

(Zhao et al., 2017, Lefebvre et al., 2013, Smailhodzic E et al., 2016). Patients mostly utilise social media sites to research a medical condition/disease, symptoms, testing, medications, and therapies. Patients can seek for and read about the experiences of other such patients. Patients get knowledge and advice from medical experts and other consumers, and then distribute it to a large number of additional consumers and clinicians via online support groups (Stewart Loane S et al., 2014).

According to studies, sufferers feel "surrounded" and accompanied by other individuals suffering from the same ailment. Patients' everyday lives have improved as a result of the emotional assistance (Clerici et al., 2012). According to research, patients prefer to seek information rather than provide information (Grajales III et al., 2014, Alsughayr A., 2015, Clerici CA et al., 2012). Regardless, patients like sharing their health-related anecdotes and experiences. Medical treatment and medication information has also been supplied. Sharing knowledge has aided other patients in developing coping techniques for their sickness (Clerici CA et al., 2012). Though many studies have found that though patients are examining, providing social media reviews and ratings of medical experts and hospitals, but just a few studies have found that people enjoy to engage with medical professionals / service providers on social media. (Moorhead et al., 2013, Van de Belt et al., 2013). On the contrary, the knowledge obtained from social media is utilised to supplement the knowledge collected from a health expert (Zhao et al., 2017, Smailhodzic et al., 2016). Furthermore, seeking for drug/medicine safety and side effects are two additional typical reasons why patients seek information from social media sites. Some utilisation include looking for information about health insurance and its providers (Fox et al., 2020, Househ et al., 2014); looking for health, food, and diet tips (Zhang et al., 2013); learning about mental health hazards (Iftikar et al., 2017); and supporting a specific health cause (Lefebvre et al., 2013, Li et al., 2018). However, there are many counter views on the use of social media platforms for health-related information. Studies like (Shoemaker et al., 2014) have observed in their study that if the medical service experts respond adversely to what patient gained from web-based media, it may diminish the patient's abstract prosperity. (Broom, 2005) has documented in his paper that doctors considered in their study considered the discussion of online media data during the meeting as a risk. Studies like (Shoemaker, Wolf & Brach, 2014) have observed that local area content gathered during clinical partnerships, made patients feel sabotaged, but on the contrary it had no effect on their Internet-based behaviour. (Nifakos et al., 2021, Kyytsonen et al., 2022) documented in their study that systematic healthcare review, recognized resistance to key social engineering attacks in keeping patient information secure. Social engineering attacks like phishing, where attempts are made to manipulate a person into disclosing personal information, for example via email (Jaeger & Eckhardt, 2021). Research findings such as (Nifakos et al., 2021) have emphasised the significance of providing uniform cybersecurity training to healthcare personnel, but have also called attention to and worry about the use of social media platforms, where personal data is occasionally exposed without adequate consideration and lack of privacy parameters leave netizens vulnerable to cyberattacks. Study like (POW, 2010) has also observed how physicians are involved in disclosing private patient information.

The present paper has mostly target to study the impact of Social media health information on Indian patients' health behaviours.

Objective

The exhaustive review of the literature related to medico-legal, technological and ethical aspects of usage of social media for medical purposes; an informed inference can be drawn that in the modern cyberspace and social media, certain patient attitudes and traits have some significant impact on usage of social media for medical purposes.

So our Research Question stands at: Does certain patient attitudes and traits have any impact on usage of social media for medical purposes?

Hypothesis

To put to test this informed guess we hypothesize the following:

$H_0$1: Medication adherence does not impact usage of social media for medical purposes.
$H_0$2: 'Unethical Practice' does not impact usage of social network for medical purposes.
$H_0$3: Self-Medication does not impact usage of social network for medical purposes.
$H_0$4: Patient-Physician relationship does not impact usage of social network for medical purposes.
$H_0$5: Belonging to marginalised patient group does not impact usage of social network for medical purposes.
$H_0$6: 'Negative behaviour' does not impact usage of social network for medical purposes.
$H_0$7: Emotional support does not impact usage of social network for medical purposes.

METHODOLOGY

This study is empirical in nature, based on patient views on certain patient attitudes and traits that may have any significant impact on usage of social network for medical purposes. Research design conclusive in nature. To collect the data here Likert scale (1-7) has been used. In the study, 1,500 persons were interviewed; this sample was judged statistically significant in the study, of which 92 did not reply;

SPSS 25 was used for all analyses, which followed a normal distribution following verification of the test results. For hypothesis testing, the 95 percent confidence threshold is utilized. The study is solely applicable to the pharmaceutical industry and only to India.

Regression Equation has been used to predict one dependent variable through seven independent variables (explained and discussed in subsequent section). The goal was to investigate the impact of certain patient attitudes and traits on usage of social network for medical purposes.

RESULT AND DISCUSSION

According to Table 1, the correlations between other variables measuring various latent constructs varied from 0.479 to 0.084. Social media marketing and self-medication are either unrelated or have a poor relationship with the other factors. The discriminant impact appears to be rather strong.

As per Table 1; the connection between social media promotion and self-medication is 0.181, indicating that they are either not associated with the other variables or have a weak association. The discriminant impact appears to be rather strong.

Table 1. Correlation matrix

Correlations									
		Self Medication	**Emotional Support**	**Marginalized Group**	**Negative Behavior**	**Unethical**	**Patient Physician Relationship**	**Medication adherence**	**Social Media Promotion**
Pearson Correlation	Self Medication	1.000							
	Emotional Support	.115	1.000						
	Marginalized Group	.149	.132	1.000					
	Negative Behavior	.129	.299	.144	1.000				
	Unethical	.109	.203	.134	.176	1.000			
	Patient Physician Relationship	.128	.122	.100	.174	.156	1.000		
	Medication adherence	.084	.148	.140	.112	.067	.111	1.000	
	Social Media Promotion	.181	.292	.291	.479	.200	.310	.093	1.000

Again, the correlation between social media promotion and emotional support is 0.292, indicating that they are either not connected with the other variables or have a weak association. The discriminant impact appears to be rather strong. Second, the correlation between social media promotion and marginalised groups is 0.291, indicating that they are either not connected with the other variables or have a weak association. The discriminant impact appears to be rather strong. Third, the association between social media promotion and bad behaviour is 0.479. This demonstrates that the relationships between them are somewhat strong. The discriminant impact appears to be rather strong. Fourth, the correlation between social media promotion and unethical behaviour is 0.200, implying that there is no relationship between them. Fifth, the association between social media promotion and patient-physician relationships is 0.310. This demonstrates that they are somewhat connected to the other factors. Finally, the correlation between social media marketing and medicine adherence is 0.093, indicating that they are either not connected with the other variables or have a weak association. The discriminant impact appears to be rather strong.

As put up in Table 2, the derived regression model explains 34.3% of the dependent variable or predictor, also this regression model with given sample size (n = 1406), with the specified number of predictors ($\Lambda = 7$) and at acceptable significance level ($\alpha = 0.05$) is found to be marginally outside the acceptable range of 1.902 – 1.922, in terms of the Durbin-Watson statistic for autocorrelation. But, since the Durbin-Watson statistic is very close to the generally accepted figure of 2.0 (for nil auto-correlation) and the sample size is large, this regression equation may be treated as acceptable in terms of the auto-correlation criteria.

As observed in Table 3, Since VIF < 4 and tolerance > 0.25, it is certain that multicollinearity issues are non-existent in the regression derived.

Moving forward, since the independent variables are found to be significant, we also could accept the unstandardized coefficient figures for construction of the regression equation.

Which is provided below in terms of *Equation 1.*

Table 2. Model summary

Model	R	R Square	Adjusted R Square	Std. Err. of the Estimate	Durbin-Watson	
					Sig. F Change	Value
1	.589[a]	.346	.343	5.74529	.000	1.964

a. Predictors: (Constant), Medication adherence, Unethical practice, Self-medication, Patient physician, Marginalized group, Negative behaviour, Emotional support

b. Dependent Variable: Socialmed

Table 3. Regression coefficients

Model		Unstandardized Coefficients		Standardized Coefficients	t	Sig.	Collinearity Statistics	
		B	Std. Error	Beta			Tolerance	VIF
1	(Constant)	3.284	1.116		2.944	.003		
	Self-medication (slfmed)	.104	.037	.062	2.801	.005	.948	1.054
	Emotional support (emospt)	.112	.022	.119	5.114	.000	.868	1.152
	Marginalized group (mrlgp)	.228	.027	.190	8.503	.000	.935	1.069
	Negative behaviour (negbeh)	.357	.023	.367	15.829	.000	.869	1.151
	Unethical practice (unethpr)	.052	.023	.050	2.227	.026	.920	1.086
	Patient physician behaviour (ptphy)	.331	.037	.199	8.911	.000	.935	1.069
	Medication adherence (medadh)	-.027	.026	-.023	-1.017	.309	.953	1.050

$$\text{Socialmed1 (pred)} = 3.284 + 0.357*(\text{negbeh}) + .331*(\text{ptphy}) + 0.228*(\text{mrlgp}) + 0.112*(\text{emospt}) + 0.104*(\text{slfmed}) + 0.052*(\text{unethpr}) - 0.027*(\text{medadh}) + e \ldots \quad (1)$$

It was found that, for every unit of increase in negative patient behaviour, there is a 0.357 increase in usage of social media for medical purposes. The term negative patient behaviour implies argumentative patients who try to influence usage of specific drugs and emphasise on being treated on specific lines of medication without substantial understanding of medical science. Additionally, they make claims of knowing about their chronic diseases and its manifestations, thus often question the diagnosis and line of treatment being executed by the doctor.

However, it was also seen that for every unit of increase in patient-physician relationship, there is a 0.331 increase in usage of social network for medical purposes. The positive patient-physician relationship implies trust and transparency between the patient and the doctor. These are interaction points where the patient acknowledges the complete authority of the doctor in treating of his/her given chronic disease and the doctor too has confidence on the drug adherence as well as general adherence to the overall medical advice provided by doctor to the patient.

For the marginalized patients, belonging from weaker sections of the society and/or belonging from geographically secluded areas or groups of people who have very little control on general awareness of their chronic diseases, it was found that every unit of increase in support for marginalized patient groups, there is a 0.228 increase in usage of social network for medical purposes.

Also, quite intuitively, it was found that for every unit of increase in emotional support to the patient (suffering from chronic diseases), there is a 0.112 increase in usage of social network for medical purposes.

Also, understandably, for every unit of increase in tendency to resort to self-medication, there is a 0.104 increase in usage of social network for medical purposes.

Worryingly true was the phenomenon that for every unit of increase in unethical behaviour, there is a 0.052 increase in usage of social network for medical purposes. Also, for every unit of increase in medication adherence, there is a 0.027 reduction in usage of social network for medical purposes.

Thus, we find that our hypotheses $H_0 1$ to $H_0 7$ stand rejected as per Equation 1 and we can safely assume that certain (predetermined/appropriate) patient attitudes and traits, do have significant impact on usage of social network for medical purposes.

From our study, it is clear that social media provides emotional support. Other researchrs have also found that (Myrick et al., 2016; Shensa et al., 2020). Social media links individuals from faraway places, including rural locations, and makes health related promotion possible (Prochaska et al., 2017). According to Latha et al. (2020), the use of social media for health by teenagers and young adults, one-quarter of the participants felt that social media may offer them with beneficial information. As per Haenssgen et al. (2021), social media has enhanced rural healthcare access. Many researchers have found that social media is having positive impact on medical adherence; but; in our study we find that there is no correlation (Cao et al., 2017; Sadun & Schanberg, 2018). Perhaps, another study require to identify the reason behind it.

People's behavioural changes as a result of mental health promotion go unreported for a long time since the outcome is intangible; this might be one of the most significant challenges of mental health promotion on social media (Moorhead et al., 2013). Due to social media lots of negative patient behaviour can be found; like; patient become argumentative, patient asking for specific diagnosis. Many researchers has identified the same outcome as us (Syed et al., 2022; Pimmer et al., 2012). Effect of social media on medicine disinformation, resulting in public bewilderment and fear, as well as increased use of self-medication, including home remedies, without demonstrated safety and efficacy. Few literature can be found that support our finding (Yang, 2020; Erku et al., 2021; Malik et al., 2020). Another issue with utilising social media is that the information published on social media is public and accessible to anybody who is interested in it. This may put responders at risk, and people's privacy may be jeopardised. There are many researchers found in their studies that the data shared by the patients might be wrongly used or credibility of the information given on social media need to be verified (Fire et al., 2019; Grajales et al., 2014). The veracity of data provided on social media platforms is another issue (Levac et al., 2010). Many feels that social media is affecting the patient physician relationship (Bosslet et al., 2011; Smailhodzic et al., 2016).

Implication

Thus, social media promotion is a two-edged sword that must be appropriately handled in order to maximise the advantages from an ethical aspect while minimizing non-constructive behavioural patterns among the target patient group (s). Further differentiation and deeper inquiry into the ethical stance as well as non-constructive behavioural qualities would be examined further in forthcoming research using relevant methodologies and instruments.

People's attitudes toward health issues, perceived benefits of action and impediments to action, and self-efficacy, according to the HBM (Health Belief Model), explain their participation (or lack thereof) in health-promoting activity (Janz & Becker, 1984; Rosenstock, 1974). In order for the health-promoting activity to be triggered, a stimulus, or cue to action, must also be present. From the study, it is clear that patients are getting different benefits as stimulus for seeking health information on social media. Same way; physicians are getting different disadvantages as stimulus which put them against SMP.

CONCLUSION

However, in the matter of usage of social media for medical purposes, there are certain significant red flag areas that need to be cleared out and minimised through concerted efforts of all concerned stakeholders in an area of social media empowerment which can otherwise liberate and benefit chronic patients to help them experience tangible medical benefits. Patients can become argumentative which might lead to wasting physicians' time. Patients might lose their trust in their physicians. Due to the influence of SMP, patients might lead to self-medication. On the other hand, it has emerged as a useful telehealth tool (especially during COVID 19). SMP provides emotional support to many patients.

If certain negative patient behavior and unethical practice are actively discouraged and adherence to doctor's medical advice is boosted/encouraged, social media has the potential to turn into the panacea that would empower patients (including marginalized and disadvantaged patients) and help them take an informed stand in their journey to be treated for their chronic diseases.

Studies associated with social media as tele-health can be done in the future. The study may be repeated with other and larger geographical regions.

REFERENCES

Al-Qahtani, M. F., Alsaffar, A. K., Alshammasi, A. R., Alsanni, G. S., Alyousef, Z. T., & Alhussaini, M. F. (2018). Social media in healthcare: advantages and challenges perceived by patients at a teaching hospital in eastern province. Saudi J Health Sciences

Alsughayr, A. R. (2015). Social media in healthcare: Uses, risks, and barriers. *Saudi Journal of Medicine & Medical Sciences*, *3*(2), 105. doi:10.4103/1658-631X.156405

Antheunis, M. L., Tates, K., & Nieboer, T. E. (2013). Patients' and health professionals' use of social media in health care: Motives, barriers and expectations. *Patient Education and Counseling*, *92*(3), 426–431. doi:10.1016/j.pec.2013.06.020 PMID:23899831

Barlett, Y. K., & Coulson, N. S. (2011). An investigation into the empowerment effects of using online support groups and how this affects health professional/patient communication. *Patient Education and Counseling*, *83*(1), 113–119. doi:10.1016/j.pec.2010.05.029 PMID:20599338

Benetoli, A., Chen, T. F., & Aslani, P. (2017). Consumer health-related activities on social media: Exploratory study. *Journal of Medical Internet Research*, *19*(10), e7656. doi:10.2196/jmir.7656 PMID:29030326

Bhaskaran, N., Kumar, M., & Janodia, M. D. (2017). Use of social media for seeking health related information–an exploratory study. *Journal of Young Pharmacists : JYP*, *9*(2), 267–271. doi:10.5530/jyp.2017.9.52

Bosslet, G. T., Torke, A. M., Hickman, S. E., Terry, C. L., & Helft, P. R. (2011). The patient–doctor relationship and online social networks: Results of a national survey. *Journal of General Internal Medicine*, *26*(10), 1168–1174. doi:10.100711606-011-1761-2 PMID:21706268

Broom, A. (2005). Virtually he@ lthy: The impact of internet use on disease experience and the doctor-patient relationship. *Qualitative Health Research*, *15*(3), 325–345. doi:10.1177/1049732304272916 PMID:15761103

Campbell, L., Evans, Y., Pumper, M., & Moreno, M. A. (2016). Social media use by physicians: A qualitative study of the new frontier of medicine. *BMC Medical Informatics and Decision Making*, *16*(1), 1–11. doi:10.118612911-016-0327-y PMID:27418201

Cao, B., Gupta, S., Wang, J., Hightow-Weidman, L. B., Muessig, K. E., Tang, W., Pan, S., Pendse, R., & Tucker, J. D. (2017). Social media interventions to promote HIV testing, linkage, adherence, and retention: Systematic review and meta-analysis. *Journal of Medical Internet Research*, *19*(11), e7997. doi:10.2196/jmir.7997 PMID:29175811

Clerici, C. A., Veneroni, L., Bisogno, G., Trapuzzano, A., & Ferrari, A. (2012). Videos on rhabdomyosarcoma on YouTube: An example of the availability of information on pediatric tumors on the web. *Journal of Pediatric Hematology/Oncology*, *34*(8), e329–e331. doi:10.1097/MPH.0b013e31825886f8 PMID:22858570

Cutrer, C. M., & Pleil, A. M. (1991). Potential outcomes associated with direct-to-consumer advertising of prescription drugs: Physicians' perspectives. *Journal of Pharmaceutical Marketing & Management*, *5*(3), 3–19.

Dixon, S. (2022). *Facebook users by country 2021*. Statista. https://www.statista.com/statistics/268136/top-15-countries-based-on-number-of-facebook-users/

Erku, D. A., Belachew, S. A., Abrha, S., Sinnollareddy, M., Thomas, J., Steadman, K. J., & Tesfaye, W. H. (2021). When fear and misinformation go viral: Pharmacists' role in deterring medication misinformation during the'infodemic'surrounding COVID-19. *Research in Social & Administrative Pharmacy*, *17*(1), 1954–1963. doi:10.1016/j.sapharm.2020.04.032 PMID:32387230

Fire, M., Goldschmidt, R., & Elovici, Y. (2014). Online social networks: Threats and solutions. *IEEE Communications Surveys and Tutorials*, *16*(4), 2019–2036. doi:10.1109/COMST.2014.2321628

Food and Drug Administration. (2020). *Generic Drugs: Questions and Answers*. FDA. https://www.fda.gov/drugs/questions-answers/generic-drugs-questions-answers

Fox, S. (2011). *The social life of health information, 2011*. California Healthcare Foundation.

Grajales, F. J. III, Sheps, S., Ho, K., Novak-Lauscher, H., & Eysenbach, G. (2014). Social media: A review and tutorial of applications in medicine and health care. *Journal of Medical Internet Research*, *16*(2), e2912. doi:10.2196/jmir.2912 PMID:24518354

Grajales, F. J. III, Sheps, S., Ho, K., Novak-Lauscher, H., & Eysenbach, G. (2014). Social media: A review and tutorial of applications in medicine and health care. *Journal of Medical Internet Research, 16*(2), e2912. doi:10.2196/jmir.2912 PMID:24518354

Greene, J. A., Choudhry, N. K., Kilabuk, E., & Shrank, W. H. (2011). Online social networking by patients with diabetes: A qualitative evaluation of communication with Facebook. *Journal of General Internal Medicine, 26*(3), 287–292. doi:10.100711606-010-1526-3 PMID:20945113

Haenssgen, M. J., Charoenboon, N., & Zanello, G. (2021). You've got a friend in me: How social networks and mobile phones facilitate healthcare access among marginalised groups in rural Thailand and Lao PDR. *World Development, 137*, 105156. doi:10.1016/j.worlddev.2020.105156

Househ, M., Borycki, E., & Kushniruk, A. (2014). Empowering patients through social media: The benefits and challenges. *Health Informatics Journal, 20*(1), 50–58. doi:10.1177/1460458213476969 PMID:24550564

Iftikhar, R., & Abaalkhail, B. (2017). Health-seeking influence reflected by online health-related messages received on social media: Cross-sectional survey. *Journal of Medical Internet Research, 19*(11), e5989. doi:10.2196/jmir.5989 PMID:29146568

India: COVID-19 impact on social media app usage 2020. (n.d.). Statista. www.statista.com. Retrieved August 1, 2022, from https://www.statista.com/statistics/1114459/india-coronavirus-impact-on-weekly-usage-time-of-social-networking-apps/

Jaeger, L., & Eckhardt, A. (2021). Eyes wide open: The role of situational information security awareness for security-related behaviour. *Information Systems Journal, 31*(3), 429–472. doi:10.1111/isj.12317

Janz, N. K., & Becker, M. H. (1984). The health belief model: A decade later. *Health Education Quarterly, 11*(1), 1–47. doi:10.1177/109019818401100101 PMID:6392204

Kyytsönen, M., Ikonen, J., Aalto, A. M., & Vehko, T. (2022). The self-assessed information security skills of the Finnish population: A regression analysis. *Computers & Security, 118*, 102732. doi:10.1016/j.cose.2022.102732

Latha, K., Meena, K. S., Pravitha, M. R., Dasgupta, M., & Chaturvedi, S. K. (2020). Effective use of social media platforms for promotion of mental health awareness. *Journal of Education and Health Promotion, 9*(1), 9. doi:10.4103/jehp.jehp_90_20 PMID:32642480

Lee, Y. C., & Wu, W. L. (2014). The effects of situated learning and health knowledge involvement on health communications. *Reproductive Health*, *11*(1), 1–7. doi:10.1186/1742-4755-11-93 PMID:25542070

Lefebvre, R. C., & Bornkessel, A. S. (2013). Digital social networks and health. *Circulation*, *127*(17), 1829–1836. doi:10.1161/CIRCULATIONAHA.112.000897 PMID:23630086

Levac, J. J., & O'Sullivan, T. (2010). Social media and its use in health promotion. Revue interdisciplinaire des sciences de la santé-Interdisciplinary. *Journal of Health Science*, *1*, 47–53.

Li, Y., & Wang, X. (2018). Seeking health information on social media: A perspective of trust, self-determination, and social support. [JOEUC]. *Journal of Organizational and End User Computing*, *30*(1), 1–22. doi:10.4018/JOEUC.2018010101

Loane, S. S., & D'Alessandro, S. (2014). Empowered and knowledgeable health consumers: The impact of online support groups on the doctor–patient relationship. [AMJ]. *Australasian Marketing Journal*, *22*(3), 238–245. doi:10.1016/j.ausmj.2014.08.007

Mackey, T. K., Cuomo, R. E., Guerra, C., & Liang, B. A. (2015). The rise of digital direct-to-consumer advertising?: Comparison of direct-to-consumer advertising expenditure trends from publicly available data sources and global policy implications. *BMC Health Services Research*, *15*(1), 236. doi:10.118612913-015-0885-1 PMID:26084705

Malik, M., Tahir, M. J., Jabbar, R., Ahmed, A., & Hussain, R. (2020). Self-medication during Covid-19 pandemic: Challenges and opportunities. *Drugs & Therapy Perspectives*, *36*(12), 565–567. doi:10.100740267-020-00785-z PMID:33041621

Markham, M. J., Gentile, D., & Graham, D. L. (2017). Social media for networking, professional development, and patient engagement. *American Society of Clinical Oncology Educational Book*, *37*(37), 782–787. doi:10.1200/EDBK_180077 PMID:28561727

Moorhead, S. A., Hazlett, D. E., Harrison, L., Carroll, J. K., Irwin, A., & Hoving, C. (2013). A new dimension of health care: Systematic review of the uses, benefits, and limitations of social media for health communication. *Journal of Medical Internet Research*, *15*(4), e1933. doi:10.2196/jmir.1933 PMID:23615206

Myrick, J. G., Holton, A. E., Himelboim, I., & Love, B. (2016). # Stupidcancer: Exploring a typology of social support and the role of emotional expression in a social media community. *Health Communication*, *31*(5), 596–605. doi:10.1080/10410236.2014.981664 PMID:26453480

Nifakos, S., Chandramouli, K., Nikolaou, C. K., Papachristou, P., Koch, S., Panaousis, E., & Bonacina, S. (2021). Influence of human factors on cyber security within healthcare organisations: A systematic review. *Sensors (Basel), 21*(15), 5119. doi:10.339021155119 PMID:34372354

Oh, H. J., Lauckner, C., Boehmer, J., Fewins-Bliss, R., & Li, K. (2013). Facebooking for health: An examination into the solicitation and effects of health-related social support on social networking sites. *Computers in Human Behavior, 29*(5), 2072–2080. doi:10.1016/j.chb.2013.04.017

Oh, H. J., & Lee, B. (2012). The effect of computer-mediated social support in online communities on patient empowerment and doctor–patient communication. *Health Communication, 27*(1), 30–41. doi:10.1080/10410236.2011.567449 PMID:21797714

Papageorgiou, A., Strigkos, M., Politou, E., Alepis, E., Solanas, A., & Patsakis, C. (2018). Security and privacy analysis of mobile health applications: The alarming state of practice. *IEEE Access : Practical Innovations, Open Solutions, 6*, 9390–9403. doi:10.1109/ACCESS.2018.2799522

Pimmer, C., Linxen, S., & Gröhbiel, U. (2012). Facebook as a learning tool? A case study on the appropriation of social network sites from mobile phones in developing countries. *British Journal of Educational Technology, 43*(5), 726–738. doi:10.1111/j.1467-8535.2012.01351.x

Pow, H. (2010). Doctors caught revealing secret patient information in Facebook posts. *Sunday Telegraph, 5*.

Prochaska, J. J., Coughlin, S. S., & Lyons, E. J. (2017). Social media and mobile technology for cancer prevention and treatment. *American Society of Clinical Oncology Educational Book, 37*(37), 128–137. doi:10.1200/EDBK_173841 PMID:28561647

Rosenstock, I. M. (1974). Historical origins of the health belief model. *Health Education Monographs, 2*(4), 328–335. doi:10.1177/109019817400200403

Sadun, R. E., & Schanberg, L. E. (2018). Using social media to promote medication adherence. *Nature Reviews. Rheumatology, 14*(8), 445–446. doi:10.103841584-018-0037-1 PMID:29915383

Shensa, A., Sidani, J. E., Escobar-Viera, C. G., Switzer, G. E., Primack, B. A., & Choukas-Bradley, S. (2020). Emotional support from social media and face-to-face relationships: Associations with depression risk among young adults. *Journal of Affective Disorders, 260*, 38–44. doi:10.1016/j.jad.2019.08.092 PMID:31493637

Shoemaker, S. J., Wolf, M. S., & Brach, C. (2014). Development of the Patient Education Materials Assessment Tool (PEMAT): A new measure of understandability and actionability for print and audiovisual patient information. *Patient Education and Counseling*, *96*(3), 395–403. doi:10.1016/j.pec.2014.05.027 PMID:24973195

Smailhodzic, E., Hooijsma, W., Boonstra, A., & Langley, D. J. (2016). Social media use in healthcare: A systematic review of effects on patients and on their relationship with healthcare professionals. *BMC Health Services Research*, *16*(1), 1–14. doi:10.118612913-016-1691-0 PMID:27562728

Smith, J. K., & Johnson, M. A. (2022). Social media promotion of branded drugs. *Journal of Pharmaceutical Marketing*, *20*(3), 123–140.

India - social media users 2023. (2020). Statista.www.statista.com › statistics › number-of-social-network-users-in-india.

Syed, S., Robinson, J., & Reddington, B. (2022). Diagnostic dilemma in psychiatry: Disease of the mind or the media? *CNS Spectrums*, *27*(2), 251–252. doi:10.1017/S1092852922000645

Van de Belt, T. H., Engelen, L. J., Berben, S. A., Teerenstra, S., Samsom, M., & Schoonhoven, L. (2013). Internet and social media for health-related information and communication in health care: Preferences of the Dutch general population. *Journal of Medical Internet Research*, *15*(10), e2607. doi:10.2196/jmir.2607 PMID:24088272

van Uden-Kraan, C., Drossaert, C., Taal, E., Seydel, E., & van de Laar, M. (2008). Self-reported differences in empowerment between lurkers and posters in online patient support groups. *Journal of Medical Internet Research*, *10*(2), e992. doi:10.2196/jmir.992 PMID:18653442

Wicks, P., Massagli, M., Frost, J., Brownstein, C., Okun, S., Vaughan, T., Bradley, R., & Heywood, J. (2010). Sharing health data for better outcomes on PatientsLikeMe. *Journal of Medical Internet Research*, *12*(2), e1549. doi:10.2196/jmir.1549 PMID:20542858

Yang, Y. (2020). Use of herbal drugs to treat COVID-19 should be with caution. *Lancet*, *395*(10238), 1689–1690. doi:10.1016/S0140-6736(20)31143-0 PMID:32422123

Zhao, Y., & Zhang, J. (2017). Consumer health information seeking in social media: A literature review. *Health Information and Libraries Journal*, *34*(4), 268–283. doi:10.1111/hir.12192 PMID:29045011

Chapter 8
Artificial Intelligence (AI) Opportunities and Challenges of Green Marketing Tourism

Wan Suryani
Universitas Medan Area, Indonesia

Azilla Luthfiyyah
Universitas Medan Area, Indonesia

ABSTRACT

Technology is experiencing a revolution that is bringing about a new era successful in changing many tasks that have been manual processes into jobs with artificial intelligence (AI) that mimic human functions in solving cognitive problems and offer the same transformative potential to replace more potential tasks and activities. Human beings in various industrial, intellectual, and social applications. This era of AI technology has surprised us with new breakthroughs in learning and decision-making and raises new challenges and opportunities for sustainable innovation in green tourism marketing. The development of AI in Indonesia is more rapid as there are many internet users. Based on the results of a survey by the Internet Service Providers Association (APJII) in the 2022-2023 period, it was recorded that there were 215.63 million people. This number increased by 2.67% compared to the previous period in 2020-2021 of 210.03 million users.

INTRODUCTION

Indonesia has various sources of income such as oil and gas. In addition to these two sectors that need more attention is the tourism sector. This sector has an important role in the development of the Indonesian economy. However, after going through

DOI: 10.4018/978-1-6684-8634-4.ch008

the pandemic some time ago, many companies in Indonesia are facing big problems to develop their business in innovative ways to remain competitive and increase their profits, one of which is a company engaged in the tourism sector. The tourism sector has in fact made a major contribution to foreign exchange earnings. Based on data from the Ministry of Tourism and Creative Economy for 2023.

Foreign exchange earnings from the Indonesian tourism sector recorded at the Ministry of Tourism and Creative Economy reached US$4.26 billion in 2022. This value increased by 769.39% compared to the previous year which was only US$0.49 billion.

According to Menparekraf Sandiaga Uno, the increase in foreign exchange from the tourism sector is inseparable from the increase in foreign tourist arrivals to Indonesia after the end of the Covid-19 pandemic.

(Hall, 2011), defines that Artificial Intelligence is intelligence where a system that shows the working conditions handled by the human mind in solving problems. Artificial intelligence (AI), (Bulchand-Gidumal, 2020) is defined as a set of abilities to understand the environment and phenomena that occur and the ability to utilize past experience and knowledge to solve new challenges appropriately (Gretzel et al., 2015). The existence of artificial intelligence today is very helpful for humans in their work and can be completed quickly and efficiently. The author defines

Table 1. Foreign exchange income from the indonesian tourism sector (2013-2023)

Artificial Intelligence as the ability of a system designed by humans to help provide information in the form of data that humans can use as a tool to solve and solve problems efficiently and effectively.

Artificial intelligence can extract large amounts of data and can be analyzed quickly and precisely. Artificial Intelligence plays into the metaverse of education enabling students to work and learn with tutors and other fellow students (Hwang & Chien, 2022). Artificial Intelligence has excelled in handling data compared to conventional ones. One of them is in understanding medical image (Sun et al., 2022). The first generation of AI research in the 1980s which produced many classical theories and classical methods, but still has computational limitations (Sun et al., 2022). There have been many studies on Artificial Intelligence that have been put forward, but no one has discussed the opportunities and challenges of Artificial Intelligence for green tourism marketing.

Talking about environmental sustainability issues and their damage. The tourism industry is one to blame for this. Due to the excessive consumption of natural resources and human behavior itself, they do not understand the importance of protecting the environment. So companies engaged in tourism must do business that is more ethical and practice it with green tourism marketing. This will make tourists who (Wan Zahari Wan Yusoff and Maziah Ismail, 2008) will be more sensitive to environmental problems (Djafarova & Rushworth, 2017).

Currently entering the era of the fourth industrial revolution where an era marked by breakthroughs in technological developments in the field of robotics, artificial intelligence (AI), internet-of-things, all of which will affect how we create and distribute value, will change the way we live, change the way we work and the way we interact (Schwab, 2016; Schulze,2019).

Evolution of Green Marketing

In the 90s, many marketing practitioners paid attention to and explored green and sustainable marketing. Existing studies start from consumer behavior, satisfaction of consumer needs, product development, product recycling and environmentally friendly packaging (Lai et al., 2013; Lewis, 2005).

Green marketing underwent a three-phase evolution (Mazzocchini & Lucarelli, 2022; Peattie, 2023).

First phase – named by Peattie the green marketing phase as ecological marketing. This phase developed in the 1960s and 1970s. The issues focus on the environment for example air pollution, depletion of oil reserves, and the environmental impact of synthetic pesticides. This phase focuses solely on industries such as automobiles, oil and chemicals.

Second phase – named by Peattie green marketing environment. This phase developed in the 1980s and 1990s. This phase developed during the Bhopal industrial tragedy in 1984, the Chernobyl incident in 1986. This phase gave rise to new ideas about sustainability and green consumers. Where people's awareness of the importance of protecting the environment has increased.

Third phase – in this phase of 2000, developing sustainable green marketing. Where many large companies began to practice environmentally friendly products. People's trust is starting to increase in its use, where previously people believed that green marketing was just advertising of products that were safe for the environment.

The Concept of Green Marketing

Green marketing is a tourism practice from increasing consumer interest in environmental sustainability (lu et al., 2013; Wan Zahari Wan Yusoff and Maziah Ismail, 2008) where the company must reconsider in implementing it into its business (Nistoreanu et al., 2020). The tendency of tourists to be interested in green practices or what we often call "environmentally friendly" began to be seen in the eighties (Martínez, 2015), where at that time sustainable economic growth in countries that experienced the development of a green environment.

Green marketing in practice refers to activities that do not have a negative impact on the environment. This concept is motivated by a natural resource-based viewpoint which refers to preventing pollution by using raw materials and energy that do not damage nature. (Kumar, 2015). The American Marketing Association defines green marketing as a process for producing, promoting or packaging products in a sustainable manner (Dean & Pacheco, 2014). Consumers are more sensitive to the environment (lu et al., 2013; Wan Zahari Wan Yusoff and Maziah Ismail, 2008) and shifting from a cost-oriented perspective to a more sustainable one (Narula & Desore, 2016). There are many advantages for companies in getting customers who are sensitive to environmental issues. Environmental orientation in terms of competitiveness, environmental values and economic objectives is a green marketing philosophy (Kumar, 2016; Dabija,2013).

(Fatoki, 2019 ; Polonsky, 1994), defines a green market as a combination of promotion and advertising of services that have a minimal impact on the environment as well as the process of manufacturing and marketing products that do not damage the environment, through specific strategies and the use of more environmentally friendly recycled materials. The author defines green tourism marketing as a combination of promotions, advertisements, services, behaviors and products that do not have a negative impact on the environment. So using an environmentally friendly marketing strategy will benefit the company because it can increase and show customer interest in developing environmentally friendly products (Líšková et al., 2016) as well as improving tourist behavior towards the environment (P. Govender & L. Govender, 2016).

Artificial Intelligence-Driven Paradigm Shift in Tourism

The marketing transformation potential that has the latest technology is AI (Verma et al., 2021) that helps organizations analyze and respond to customers (Wirth, 2018). Where AI is the most suitable solution for marketing functions for practitioners around the world that offers consumer insights about behavior for customer attraction and retention. Much literature has highlighted the importance of AI in marketing and charted future research directions (Verma et al., 2021). AI has provided digital solutions to attract and retain customers (Anshari et al., 2019; Bolton et al., 2018), AI provides a competitive advantage (Rouhani et al., 2016; Spring et al., 2017), and can facilitate product offerings and customer service (Balaji & Roy, 2017; Khanagha et al., 2017)(Balaji & Roy, 2017; Khanagha et al., 2017; Liao, 2015).

The Important of Artificial Intelligences in Green Marketing Tourism

AI is a new technology that can be used in the tourism industry that can help provide new experiences to customers. Technologies in AI include facial recognition technology, virtual applications, chatbots, robots, google maps, language translators, and others. (Samala et al., 2022).

AI can be used in the field of green tourism marketing to create robots or mobile applications that can connect or interact directly with their environment. AI also helps companies engaged in green tourism marketing in extracting information from data and helps find patterns and correlations between various data, identify existing trends, and predict future behavior (Verma et al., 2021). In a business, AI can also help improve business processes by optimizing decision making and helping to increase productivity. AI technology is trained to learn more than just a specific task (Verma et al., 2021).

The Opportunities AI in Green Marketing Tourism

AI can surprise us with new breakthroughs in learning and decision-making and pose new challenges and opportunities for sustainable innovation in green tourism marketing. AI can be transferred to machines to carry out tasks from the simplest to the most complex (Verma et al., 2021).

AI can provide opportunities for green tourism marketing in the following ways:

1. 1. Creating a better visitor experience: AI can help improve customer experience in green tourism marketing by giving them access to relevant and appropriate information and helping customers make informed decisions. AI can also help improve the green travel customer experience by identifying and suggesting the right eco-friendly services and products according to the needs of green travel customers.
2. Help build awareness: Artificial awareness (AI) can help promote green tourism marketing by building awareness about destinations, eco-friendly products and services provided by companies.
3. Use of analytic algorithms: Artificial awareness (AI) helps collect data about visitors, green tourism marketing content, and the types of green tourism services they are looking for. All of this will help make it easier for stakeholders to optimize the experience felt by visitors. AI also helps increase return on investment by providing trending information about green tourism marketing and ensuring that what customers perceive is the best experience and makes for an interesting trip.

The Challenges AI In Green Marketing Tourism

AI is paving the way for a green tourism industry and is able to reduce manual processes. The travel industry has always been a major pioneer in adopting technology, and companies operating in the travel industry are welcoming AI with open arms (World Al Show, 2019). The tourism industry uses AI as a destination to increase the level of personalization, adapt customer recommendations and guarantee fast response times even when workers are not present. AI is becoming so important in the travel industry setting that it is used to assist and communicate with customers and maintain engaged customer relationships (Samala et al., 2022; Revfine.com,2019).

Even though AI is developing so rapidly, there are still limits to its use for those who are illiterate (Reddy, 2006). This limitation must have innovative solutions coupled with technology that is able to solve these problems. The fact that "artificial intelligence" provides benefits to the industrial world, especially to the tourism industry in terms of monetary benefits by replacing employees and providing nonmonetary benefits by providing new experiences to customers, but all of that still cannot go beyond human intelligence (Laurent et al., 2015).

Another big challenge for tourism marketing in terms of using AI is data safety and security. This is related to the challenges in implementing AI in several important sectors such as finance, visitor travel history which is related to privacy and visitor data security issues. (kannan dan Bernoff,2019).

Apart from the challenges of using Artificial Intelligence above, there are also challenges that must be faced by tourism companies in implementing green tourism marketing. It comes from the environment around the tourist destination. Where there is still a lack of understanding of some people in protecting the environment and understanding about protecting the environment is a shared responsibility and not only the responsibility of the government and the tourism industry. All of this is a collective homework, be it the government, the tourism industry, business actors and all levels of society.

METHODOLOGY

This chapter uses content analysis to elaborate or explain the opportunities and challenges of AI in green marketing tourism. Green tourism marketing conceptualizes sustainable tourism development, which focuses on two things: first, sustainable tourism as an economic activity. Second, tourism as an element of a broader sustainable development policy and in line with the principles of sustainable development.

CONCLUSION

As technology develops, the presence of Artificial Intelligence is increasingly helpful in providing many benefits in green tourism marketing. This is marked by the increasingly widespread application of innovative technologies in which the tourism industry will achieve maximum profits in the future. This technology will grow by more than 9 percent between 2010 and 2023 (Samala et al., 2022). The main factors that provide expected benefits from the performance of tourist trips are from the tourism industry.

AI can surprise the tourism industry with new breakthroughs in learning and decision-making and pose new challenges and opportunities for sustainable innovation in green tourism marketing. AI can be transferred to machines to carry out tasks from the simplest to the most complex (Verma et al., 2021).

AI is paving the way for a green tourism industry and is able to reduce manual processes. The travel industry has always been a major pioneer in adopting technology, and companies operating in the travel industry are welcoming AI with open arms.

In this paper, what should not be forgotten is giving the community the opportunity to manage tourism objects properly and professionally. Communities must be involved so that indirectly they feel responsible for preserving tourism and local culture, not just using tourist objects as a source of income. This understanding is important to be built by all green tourism stakeholders throughout Indonesia.

REFERENCES

Anshari, M., Almunawar, M. N., Lim, S. A., & Al-Mudimigh, A. (2019). Customer relationship management and big data enabled: Personalization & customization of services. *Applied Computing and Informatics*, *15*(2), 94–101. doi:10.1016/j.aci.2018.05.004

Balaji, M. S., & Roy, S. K. (2017). Value co-creation with Internet of things technology in the retail industry. *Journal of Marketing Management*, *33*(1–2), 7–31. doi:10.1080/0267257X.2016.1217914

Bolton, R. N., McColl-Kennedy, J. R., Cheung, L., Gallan, A., Orsingher, C., Witell, L., & Zaki, M. (2018). Customer experience challenges: Bringing together digital, physical and social realms. *Journal of Service Management*, *29*(5), 776–808. doi:10.1108/JOSM-04-2018-0113

Bulchand-Gidumal, J. (2020). Impact of Artificial Intelligence in Travel, Tourism, and Hospitality. Handbook of E-Tourism, 1–20. doi:10.1007/978-3-030-05324-6_110-1

Dean, T. J., & Pacheco, D. F. (2014). Green marketing: A strategic balancing act for creating value. *The Journal of Business Strategy*, *35*(5), 14–22. doi:10.1108/JBS-11-2013-0109

Djafarova, E., & Rushworth, C. (2017). Exploring the credibility of online celebrities' Instagram profiles in influencing the purchase decisions of young female users. *Computers in Human Behavior*, *68*, 1–7. doi:10.1016/j.chb.2016.11.009

Fatoki, O. (2019). Green Marketing Orientation and Environmental and Social Performance of Hospitality Firms in South Africa. *Foundations of Management*, *11*(1), 277–290. doi:10.2478/fman-2019-0023

Govender, J., & L. Govender, T. (2016). The influence of green marketing on consumer purchase behavior. *Environment and Ecology*, *7*(2), 77–85. doi:10.21511/ee.07(2).2016.8

Gretzel, U., Sigala, M., Xiang, Z., & Koo, C. (2015). Smart tourism: Foundations and developments. *Electronic Markets*, *25*(3), 179–188. doi:10.100712525-015-0196-8

Hall, C. M. (2011). Health and medical tourism: A kill or cure for global public health? *Tourism Review*, *66*(1/2), 4–15. doi:10.1108/16605371111127198

Hwang, G. J., & Chien, S. Y. (2022). Definition, roles, and potential research issues of the metaverse in education: An artificial intelligence perspective. In *Computers and Education* (Vol. 3). Artificial Intelligence., doi:10.1016/j.caeai.2022.100082

Khanagha, S., Volberda, H., & Oshri, I. (2017). Customer Co-Creation and Exploration of Emerging Technologies: The Mediating Role of Managerial Attention and Initiatives. *Long Range Planning*, *50*(2), 221–242. doi:10.1016/j.lrp.2015.12.019

Kumar, P. (2015). Green marketing innovations in small Indian firms. *World Journal of Entrepreneurship, Management and Sustainable Development*, *11*(3), 176–190. doi:10.1108/WJEMSD-01-2015-0003

Kumar, P. (2016). State of green marketing research over 25 years (1990-2014): Literature survey and classification. *Marketing Intelligence & Planning*, *34*(1), 137–158. doi:10.1108/MIP-03-2015-0061

Lai, K. H., Wu, S. J., & Wong, C. W. Y. (2013). Did reverse logistics practices hit the triple bottom line of Chinese manufacturers? *International Journal of Production Economics*, *146*(1), 106–117. doi:10.1016/j.ijpe.2013.03.005

Laurent, P., Chollet, T., & Herzberg, E. (2015). *Intelligent automation entering the business world.* Deloitte. https://www2.deloitte.com/content/dam/Deloitte/lu/Documents/operatihttps://www2.deloitte.com/content/dam/Deloitte/no/Documents/technology/intelligent-automation-entering-the-business-world.pdf

Lewis, H. (2005). Defining product stewardship and sustainability in the Australian packaging industry. *Environmental Science & Policy*, *8*(1), 45–55. doi:10.1016/j.envsci.2004.09.002

Liao, T. (2015). Augmented or admented reality? The influence of marketing on augmented reality technologies. *Information Communication and Society*, *18*(3), 310–326. doi:10.1080/1369118X.2014.989252

Líšková, Z. D., Cudlínová, E., Pártlová, P., & Petr, D. (2016). Importance of Green Marketing and Its Potential. *Visegrad Journal on Bioeconomy and Sustainable Development*, *5*(2), 61–64. doi:10.1515/vjbsd-2016-0012

lu, L., Bock, D., & Joseph, M. (2013). Green marketing: What the Millennials buy. *Journal of Business Strategy, 34*(6), 3–10. doi:10.1108/JBS-05-2013-0036

Martínez, P. (2015). Customer loyalty: Exploring its antecedents from a green marketing perspective. *International Journal of Contemporary Hospitality Management*, *27*(5), 896–917. doi:10.1108/IJCHM-03-2014-0115

Mazzocchini, F. J., & Lucarelli, C. (2022). Success or failure in equity crowdfunding? A systematic literature review and research perspectives. *Management Research Review*. doi:10.1108/MRR-09-2021-0672

Narula, S. A., & Desore, A. (2016). Framing green consumer behaviour research: Opportunities and challenges. *Social Responsibility Journal*, *12*(1), 1–22. doi:10.1108/SRJ-08-2014-0112

Nistoreanu, P., Aluculesei, A. C., & Avram, D. (2020). Is green marketing a label for ecotourism? The Romanian experience. *Information (Basel)*, *11*(8), 1–13. doi:10.3390/info11080389

Peattie, K. (2023). *Towards sustainability : The third age of green marketing*. Research Gate.

Polonsky, M. J. (1994). An introduction to green marketing. *Electronic Green Journal*, *1*(2). doi:10.5070/G31210177

Reddy, R. (2006). in Support of Society. *IEEE Intelligent Systems*, *21*(3).

Rouhani, S., Ashrafi, A., Zare Ravasan, A., & Afshari, S. (2016). The impact model of business intelligence on decision support and organizational benefits. *Journal of Enterprise Information Management*, *29*(1), 19–50. doi:10.1108/JEIM-12-2014-0126

Samala, N., Katkam, B. S., Bellamkonda, R. S., & Rodriguez, R. V. (2022). Impact of AI and robotics in the tourism sector: A critical insight. *Journal of Tourism Futures*, *8*(1), 73–87. doi:10.1108/JTF-07-2019-0065

Schwab, K. (2016). The Fourth Industrial Revolution: what it means and how to respond. *World Economic Forum*, 1–7.

Spring, M., Hughes, A., Mason, K., & McCaffrey, P. (2017). Creating the competitive edge: A new relationship between operations management and industrial policy. *Journal of Operations Management*, *49–51*(1), 6–19. doi:10.1016/j.jom.2016.12.003

Sun, Z., Sandoval, L., Crystal-Ornelas, R., Mousavi, S. M., Wang, J., Lin, C., Cristea, N., Tong, D., Carande, W. H., Ma, X., Rao, Y., Bednar, J. A., Tan, A., Wang, J., Purushotham, S., Gill, T. E., Chastang, J., Howard, D., Holt, B., & John, A. (2022). A review of Earth Artificial Intelligence. In Computers and Geosciences, 159. doi:10.1016/j.cageo.2022.105034

Verma, S., Sharma, R., Deb, S., & Maitra, D. (2021). Artificial intelligence in marketing: Systematic review and future research direction. *International Journal of Information Management Data Insights*, *1*(1), 100002. doi:10.1016/j.jjimei.2020.100002

Wan, Z. W. Y., & Ismail, M. (2008). 기사 (Article) 와 안내문 (Information). *The Electronic Library*, *34*(1), 1–5.

Wirth, N. (2018). Hello marketing, what can artificial intelligence help you with? *International Journal of Market Research*, *60*(5), 435–438. doi:10.1177/1470785318776841

Chapter 9
A Study on Smart Energy Management Framework Using Cloud Computing

N. Venkateswaran
Panimalar Engineering College, India

Sahil Manoj Chavan
Sandip University, India

K. Vidhya
Karunya Institute of Technology and Sciences, India

K. Sekar
Hindusthan College of Engineering and Technology, India

Manimuthu Ayyannan
Independent Researcher, USA

Sampath Boopathi
Muthayammal Engineering College, India

ABSTRACT

This study focuses on investigating energy-aware resources and designing an energy-efficient cloud architecture. It aims to optimize the decentralization model and establish migration behavior to improve IT business operations using cloud computing. Several models were evaluated for performance metrics such as physical host use, power utilization, and service level agreement (SLA) misuse. These models include smart cloud optimization for resource configuration handling (SCORCH), V-MAN, decentralized energy-aware collaborative model (DEACM), efficient energy-aware resource management model (EEARMM), and energy and power-aware dynamic migration (EPADM). To handle resource configuration efficiently, the study employed V-MAN, DEACM, EEARMM, and EPADM within the framework of smart cloud optimization. Among these models, EPADM proved to be the most effective for resource utilization through dynamic VM migration. By leveraging EPADM, the cloud system can dynamically migrate virtual machines (VMs) to optimize resource allocation and energy consumption.

DOI: 10.4018/978-1-6684-8634-4.ch009

INTRODUCTION

Cloud computing is a transformative technology in IT infrastructure and resource management services. It operates on the XaaS model, offering various services like Software as a Service (SaaS), Platform as a Service (PaaS), and Infrastructure as a Service (IaaS). Cloud computing's pay-per-use model allows users to pay for resources based on their specific needs, ensuring Quality of Service (QoS) and privacy protection.

Leading companies in the industry, such as Microsoft, Google, and IBM, have invested heavily in building data centers to provide cloud computing services. These data centers serve as the backbone of cloud infrastructure, hosting a wide range of tools, programs, and application development frameworks. Through cloud computing, customers can access these resources using a web browser, eliminating the need for local infrastructure and enabling remote access to computational services. The underlying principle of cloud computing is resource sharing among remote computers. By pooling and efficiently allocating computational resources, cloud providers ensure that businesses can access the necessary resources on-demand. The payment mechanism in cloud computing is closely tied to resource utilization, allowing customers to pay based on their actual usage rather than a fixed subscription fee(Boopathi, Siva Kumar, et al., 2023; Boopathi, Venkatesan, et al., 2023; Reddy et al., 2023; Sampath et al., 2022).

Quality of Service (QoS) is a critical aspect of cloud computing. It encompasses factors such as performance, reliability, availability, and security, ensuring that cloud services meet the expectations and requirements of users. Privacy protection is another significant concern, as cloud computing involves the storage and processing of sensitive data. Cloud providers must implement robust security measures and adhere to privacy regulations to safeguard user information. In summary, cloud computing has revolutionized the IT industry by providing dynamic resource provisioning through a variety of services. By leveraging the benefits of cloud infrastructure, businesses can access computational resources on-demand, while the pay-per-use model ensures efficient resource utilization. Quality of Service (QoS) and privacy protection are vital considerations in the design and operation of cloud computing systems, ensuring reliable and secure service delivery to users (Alhasnawi et al., 2020; Sheikhi et al., 2015a).

Cloud computing has witnessed a significant migration of customers due to its numerous advantages. It offers online services delivered over the Internet, enabling users to access a wide range of capabilities and resources. One of the key aspects of cloud computing is cloud storage services, which provide on-demand computing capabilities, flexible pay-as-you-go pricing models, and the ability to host multiple users on shared physical infrastructure. Cloud storage services empower consumers with virtually unlimited computational power. Platforms like Google Drive, Google Photos, Amazon Cloud, and others allow users to rent computing resources on an hourly basis. These services cater to various needs, including online file backup, computer picture editing, digital video tools, integration with social media platforms like Twitter, image album creation, web antivirus, application processing, word processing, travel and management software, and computer-based e-commerce solutions. Cloud computing is built on a virtualized infrastructure that combines data computation, storage, and software resources into a unified network and server environment. It leverages technologies such as clustering, grid computing, and cloud computing to aggregate resources and provide a seamless and unified view of the system. This unified picture allows users to access and utilize a diverse set of computing resources without having to manage the underlying infrastructure. By embracing cloud computing, businesses and individuals can benefit from the scalability, flexibility, and cost-effectiveness it offers. The ability to access a wide range of online services and resources provides users with unprecedented convenience and productivity. Cloud computing continues to evolve and innovate, driving digital transformation and enabling new possibilities in various industries (Luján et al., 2019; Yigit et al., 2014).

Cloud computing has become a prominent paradigm for delivering internet services, offering both individuals and businesses cost savings and convenience. It has found applications in various sectors such as banking, e-commerce, retail, and education, enabling organizations to minimize capital expenditure on IT infrastructure. The popularity of cloud computing continues to grow as the demand for services increases. However, this growth also leads to rising processing costs and the expansion of cloud data centers, which consume significant amounts of energy. In order to ensure high levels of customer satisfaction, it is crucial for cloud providers to meet Service Level Agreement (SLA) requirements while minimizing energy consumption(Anitha et al., 2023; Babu et al., 2023; Boopathi, Arigela, et al., 2023; Jeevanantham et al., 2023; Subha et al., 2023).

Meeting SLA requirements and optimizing energy consumption pose significant challenges for cloud providers. They need to invest in robust cloud management and coordination tools to efficiently allocate resources, optimize workloads, and maintain performance levels. These tools help providers monitor and control resource usage, ensure reliability, and deliver on customer expectations while minimizing energy waste. It is not just large enterprises that can benefit from cloud management and coordination tools. Small and medium-sized businesses can also leverage these tools to achieve various objectives. They can utilize cloud management platforms to streamline their IT operations, enhance efficiency, and reduce costs. These platforms offer features such as resource monitoring, workload management, and scalability, enabling businesses to leverage cloud services effectively and meet their specific needs.

By adopting robust cloud management and coordination tools, cloud providers can enhance their service offerings and meet SLA requirements while optimizing energy consumption. Simultaneously, businesses of all sizes can take advantage of these tools to leverage the benefits of cloud computing and improve their operations. As the demand for cloud services continues to rise, it becomes imperative for both providers and businesses to invest in effective management and coordination solutions to ensure optimal performance, customer satisfaction, and sustainability in the cloud computing ecosystem (Alhasnawi et al., 2021; Chang et al., 2021).

In the cloud environment, workload management among virtual machines (VMs) is a fundamental paradigm for ensuring efficient resource utilization and meeting Service Level Agreements (SLAs). Infrastructure providers strive to distribute workloads effectively, ensuring that different users' programs are allocated resources independently of each other within the shared physical infrastructure. However, effectively monitoring the resource allocation mechanism and making accurate decisions can be challenging. Virtual machine migration presents an opportunity to address performance issues faced by Cloud Service Providers (CSPs). By dynamically migrating VMs between physical hosts, CSPs can optimize resource utilization, balance workloads, and improve overall system performance. VM migration allows for workload consolidation, reducing underutilization and improving resource efficiency. Additionally, green computing techniques are employed to reduce energy consumption in large-scale cloud deployments. Energy-efficient practices and technologies are adopted to minimize the environmental impact and operational costs associated with running cloud services. However, the consolidation of networks in cloud environments can introduce instability due to challenges in resource management.

Efficient resource management is crucial in achieving stability and maximizing the benefits of network consolidation. Monitoring and decision-making mechanisms need to be in place to ensure that resources are allocated optimally, avoiding bottlenecks and performance degradation. This involves continuously monitoring resource usage, predicting workload patterns, and making informed decisions regarding workload distribution and VM migration. By effectively managing resources and leveraging techniques such as VM migration and green computing, CSPs can address performance issues, improve energy efficiency, and enhance the overall quality of cloud services. However, it is important to carefully design and implement resource management strategies to maintain stability and meet SLA requirements while minimizing energy consumption in a consolidated network environment.

Need for Cloud Computing: IT necessitates a high investment cost, frequent resource changes, and operational and maintenance costs, making transaction processing challenging. Cloud computing is more efficient than previous techniques since the user may access the cloud and use the resource. When developing cloud-based resource management concepts, third-party cloud supplies and security concerns must be addressed. Cloud computing offers a number of qualities that must be applied to study findings, such as approximated service, resource polling, wide network access, on-demand self-service, and quick flexibility. Cloud data has various dimensions, and typical big data applications with a mining method for privacy are independent sources and decentralized control. Without depending on coordinated control, data sources are leveraged to produce and aggregate messages(Allahvirdizadeh et al., 2019; Hashmi et al., 2021).

Virtual Data Centre Framework: Initially used to host public-facing apps, businesses began to migrate internal line-of-business systems to the cloud to improve security, stability, efficiency, and cost. Additional infrastructure and networking services were constructed to enable this reliability, as well as new capability for elastic scale, disaster recovery, and other considerations.

Cloud Service Providers (CSPs) must be adaptive in order to offer applications that satisfy client requirements while consuming the least amount of energy and cash. To ensure the effectiveness of a computational model, virtualization strategies such as resource usage, scheduling, and mapping are required. This can help data centers.

OBJECTIVES

To enhance load balancing and enable cloud services for effective resource management, cloud resource management rules may be classified into five types. The suggested research work's major goals are to improve people's quality of life.

- This study focuses on enhancing the energy efficiency of cloud architecture by considering energy-conscious resources. It aims to develop compute energy-aware equations and establish a decentralization optimization model. Additionally, the study aims to determine the optimal workload vector for virtual machine migrations.
- To achieve these objectives, the study introduces the Effective Energy-Aware Resource Management (EEARM) model, which facilitates the movement of virtual machines based on energy considerations. The Energy and Power-Aware Dynamic Migration (EPADM) technique is employed to ensure reliable and low-energy operation in the cloud environment.
- The performance of virtual machine migrations is evaluated based on key parameters such as physical host utilization, power consumption, and

adherence to Service Level Agreements (SLAs). Several resource configuration management approaches, namely SCORCH, V-MAN, DEACM, EEARMM, and EPADM, are compared to assess their effectiveness in optimizing these performance metrics.

- The ultimate goal of this research is to improve supplier dependability and deliver excellent customer service in the cloud computing domain. Accuracy plays a crucial role, as the algorithms used should provide outputs that maximize utility and closely align with the desired objectives. By optimizing energy efficiency and resource management, cloud providers can enhance their service reliability and ensure customer satisfaction.

This study proposes an effective resource distribution using VMs in the cloud, utilizing VM's complicated load balancing mechanism and fine-tuning it by concentrating on energy usage and workload mapping. Efficient Energy-Aware Resource Management Model and Decentralized Energy-Aware Collaborative Model. To ensure Quality of Service (QoS) and decrease energy usage, cloud infrastructures employ DEACM, Target-based VM Migration, EPADM, Estimation Model Algorithm, and EEARM. DEACM is used to properly balance load, EPADM is utilized to efficiently map resources, and EEARM enables efficient resource use.

DEACM is a virtualization technology used to balance load and energy utilization and assure service efficiency in a cloud computing system. To prevent workload variations, initialization and architectural analysis are examined. To disperse nodes, Local Index Transmission is employed. Energy-Aware and Optimization for Cloud Service Providers (EEARMM) was created to investigate the migration of virtual machines to balance the load. Physical machine (PM) and memory allocation breach are two factors. Workload prioritization, workload balance, and the selection method are used to determine VM migrations.

DEACM

In cloud computing, resource management is a crucial and difficult competition. It entails the acquisition, utilization, and autonomous control of a shared resource. Quality of Service is one of the most serious issues (QoS). The QoS-conscious resource optimization challenge is a viable technique to reach an objective such as lowest throughput, cost, and response time. The purpose of provisioning is to assign the optimal resource to the intended recipient depending on customer need. The analysis of resources is used to locate and download current services based on attributes maintained in cloud storage providers. To optimally map resources, an optimized decentralized collaborative prototype was presented utilizing the Energy-Aware Model(Khayyam et al., 2013; Naveen et al., 2016).

Cloud storage infrastructures give large-scale data centers to customer-side applications, and virtualization technologies have been adopted to multiplex important sources. To maintain cloud QoS, resource allocation strategies must be managed. The DEACM algorithm is a power optimization technique that distributes resources among users. Cloud computing refers to a network that provides a large capacity pool and allows data to be stored in a virtual, limitless storage space system. Service providers have assisted customers in storing data on the cloud(Boopathi et al., 2021; Boopathi & Myilsamy, 2021; Vignesh et al., 2018).

Large-Scale Data Centres in Cloud

To properly run large-scale data centers and utility clouds, operators must consider existing infrastructure and device patterns. This involves continuous real-time tracking and online analysis of data gathered by monitoring systems and analytics. Costs and delays must be balanced against the benefits of quickly identifying and responding to unfavourable or non-performing device situations. Figure 1 represents the large-scale data centre cloud architecture. Virtual machines are used to run or stop actual computers dynamically based on workloads and resources. Virtual machines can execute multiple applications on a distinct PM, allowing cloud users with accessibility and elasticity. Depending on the physical archetype, services are obtained through several VMs. Migration of VMs across PMs is required to combine variable loads and deliver optimal services(Domakonda et al., 2023; Kumara et al., 2023; Reddy et al., 2023; Selvakumar et al., 2023; Vennila et al., 2023).

Figure 1. Large Scale Cloud Data Centre Architecture

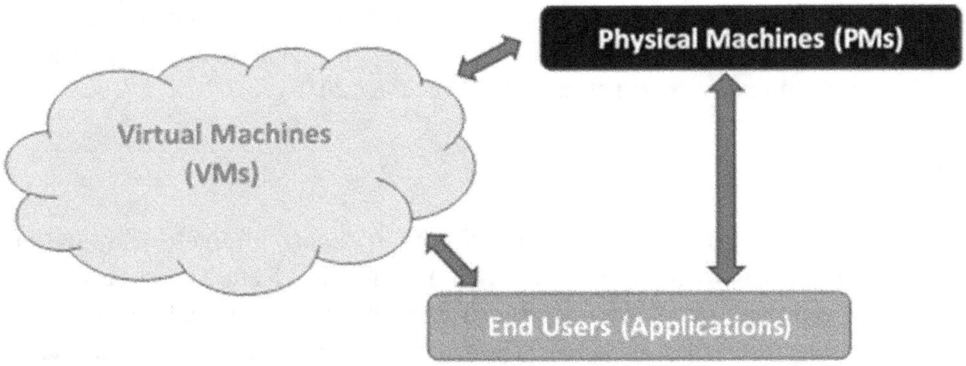

Energy-Aware Model

The DEACM is a group paradigm that focuses on CPU energy and power consumption. To reduce energy usage, Dynamic Voltage and Frequency Scaling (DVFS) has been adopted. The utilization of CPU resources is directly proportional to the workload rate, and the PM's energy consumption is dictated by the CPU resource utilization needs. To offer usability certification, a decentralized, optimized paradigm for virtual machine administration is required. The active nodes in the local index transfer model between machines conduct the forwarding procedure at predefined time intervals.

Collaborative Algorithm

Target-based VM Migration: Algorithmic descriptions of VM migration and resource planning are provided.

Workload Vector Calculation: PMs must acquire load knowledge in order to make VM migration choices, and each PM controls a load vector that accepts load indexes from others.

VM Selection Process

The two-level threshold estimation approach is used to reduce energy consumption by shifting certain VMs to other PMs when they reach a specified threshold value. The Minor Threshold (MnT) number is utilized to keep excess processor power available to accelerate workloads and meet SLAs, whereas the Major Threshold (MrT) value is used to switch on more PMs that are not totally dedicated to sleeping. Before the VMs reach a state where their current usage is less than the predefined major threshold value, they must be checked. If the actual consumption exceeds MrT, the RC assessment should be less than MrT but larger than the limiting threshold rate. Virtual machines are distributed until the RC value falls below the primary threshold point. If the resource utilization of the PM is less than the lower threshold number, it should be designated as the minimum used, and all VMs that rely on a PM will be chosen for migration(Farid et al., 2021; Ye et al., 2019).

Procedure

The next stage in virtual machine migration is to choose a target for migration. For the Best Match protocol, the Select Target (vmToMn) feature must be utilized. The optimized method is shown in Bench 2. Before pursuing goals, it is critical to explain the virtual machine's RC. The machined source in vmToMn is used to wake up sleeping PMs and begin the migration process. If an error occurs, the goal finding process is halted, and vmToMn returns a migration decision based on the most recent PM. A virtual migration idea based on two-level threshold estimate helps manage loads across physical devices in a large-scale cloud data center.

Results and Analysis

To illustrate its practicality, the proposed Decentralized Energy-Aware Collaborative Model (DEACM) is analyzed using the CloudSim tool and compared to current models such as SCORCH and V-MAN. A decentralized methodology was utilized in the study to establish pricing for two-level thresholds, minor and major threshold values. Using the two-level threshold method in the migration VM selection process, the current DEACM is predicted to achieve a minimum energy usage of 25%. This enables underutilized devices to enter a sleeping state, which uses less energy than the active mode. The graph indicates that underutilized nodes have a greater energy consumption at the start of the simulation, but as time passes, they are freed and collapse into the sleeping node community, lowering energy consumption.

The study offers a method for selecting eligible VMs for migration. It also considers computers that are idle, underutilized, or overutilized. The V-MAN and SCORCH models define a resource management method that is similar across virtual machines. Instead of overwhelming the existing compute machines, the proposed DEACM decided to activate additional compute machines. The V-MAN and SCORCH models establish a common method for resource management among virtual machines, but the suggested DEACM initiates new computing devices rather than overloading existing ones. It also considers machines that are idle, underutilized, or overutilized. The suggested work measures energy usage and develops a VM movement strategy that provides better results and outperforms others(Bera et al., 2015; Sittón-Candanedo et al., 2019).

EEARMM

Cloud computing has become an important aspect of the IT industry, making software as a service more desirable and influencing how IT gear is developed and purchased. Virtualization technology is a key component of current cloud computing designs, and data centers are employed to provide cloud services. This necessitates the development of solutions that minimize operational expenses and power consumption. Server virtualization is the most common technique used in cloud data centers, where numerous virtual machines (VMs) run concurrently on a single Physical Machine (PM) or host server (Figure 2). Each VM runs a guest

Figure 2. EEARMM model flow chart

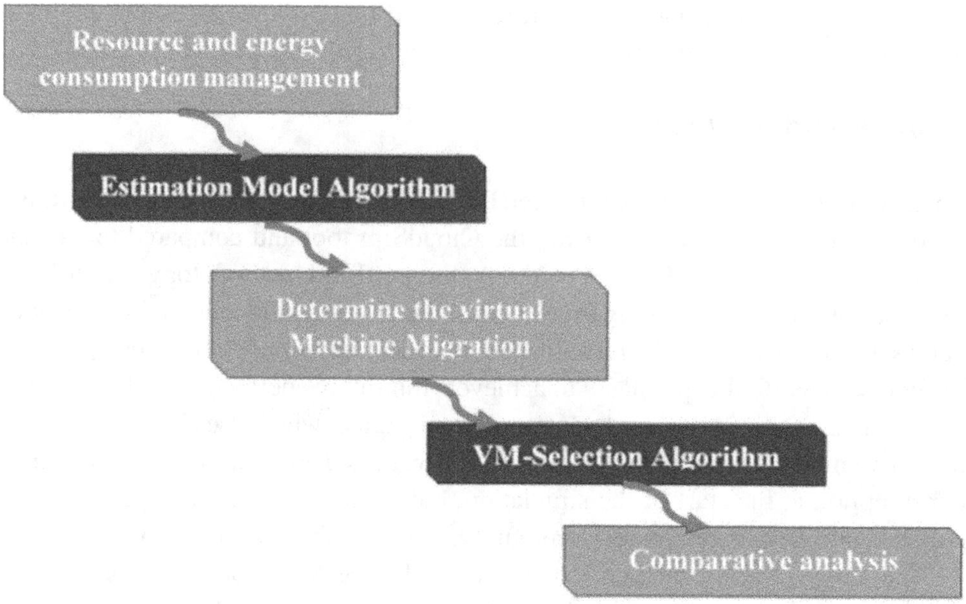

operating system, software, and applications and allocates resources from a pool of available resources.

The key issues are the placement of VMs in appropriate PMs and the necessity to transfer them between PMs. In the cloud, effective resource control dictates the allocation of computing services to customer requests in order to meet user requirements. However, in an impulsive cloud domain, CSPs and end users are difficult to persuade. This chapter describes an efficient model of energy-aware resource management (EEARMM). The prototype's goal is to decrease the number of migrations required for successful resource utilization through precise workload control. It determines virtual machine migration using the Estimation Model Algorithm (EMA) and the VM-Selection Algorithm (SA), preventing overloading and enhancing energy efficiency(Reka & Ramesh, 2016; Zhang et al., 2016). The expected prototypical's efficacy, likelihood, and scalability in cloud dominion are demonstrated by performance evaluation and comparison study. To overcome the drawbacks of prior topologies, this chapter presents a decentralized, energy-efficient workload balancing architecture for cloud infrastructures. It manages converted VM cases by mapping virtual machines to actual machines in a strong overlay cloud network. The distributed paradigm necessitates the involvement of physical devices in load balancing in the absence of centralized administration or modules. To decrease migrations, estimation models for resource and energy consumption are employed, virtual machine migration is decided by the VM-Selection method, and task allocations and resource optimization are performed using VM collection and relocation strategies(Khajenasiri et al., 2017; Sheikhi et al., 2015b). The amount of communicative knowledge transferred reduces the number of links that systems can keep. Powerful deployment approaches in large-scale cloud systems have frequently reduced capital consumption. Biased VM migration methods have effectively overcome bottleneck and connection failure concerns(Boopathi, 2019, 2021, 2022b, 2022a, 2022c).

Decentralization: Using hierarchical algorithms, the resource management paradigm is disseminated and independently tested.

Flexibility: In a cloud data center, live VM migration is utilized to improve model performance in the utilization of energy resources.

Feasibility: In the cloud model, cost efficiency is important, and the EEARMM implements complex VM migration to reduce the number of active physical devices, thereby reducing the number of maximum physical devices consumed.

Dynamic VM Migration Model

The suggested approach focuses on minimizing the number of physical machines in operation, balancing task control, and distributing resources effectively for each PM.

Selection Process

Due to resource limits, EEARMM is a discovery technique for selecting suitable virtual machines to migrate from numerous computers, which would have a significant impact on overall processing in the cloud data center. The dynamic resource utilization approach is used to choose eligible VMs for migration. To choose the next VM, the program examines data history and employs assessment methodologies. For more frequent memory use, the Most Active PM is discovered in the list of VM hosts, and the virtual computer with the lowest energy consumption or resource utilization is regarded the best option for migration.

Selection of Host: In the event of a new Most Active PM in the cloud, it is critical to set the cloud migration role to transfer a VM to a physical device with a higher next periodic load, or to pick an appropriate destination host. The destination host is chosen based on the following criteria: Minimum Workload Priority, Workload Balancing, and Selection Algorithm.

Workload Priority: Prioritization-based migration to physical devices with low energy consumption assures the VM's effective execution, but does not account for resource utilization improvements over time.

Workload Balance: The idea is to have the VM process operate in numerous PMs, which will improve overall speed.

Algorithm: The monitor segment, which comprises of numerous dispersed virtual machines in a cloud system, gathers data histories on distinct PMs' memory use. It considers resource use, minimizing the number of VM migrations required in corporate clouds. It is critical to select the appropriate VM and location for migration.

METHODOLOGY

To evaluate the parameters of the prototype model, which accounts for complex variation in VM memory utilization, the Least-Square estimation model was utilized.

EEARMM requires initializing mandatory parameters to apply.

- The resource monitor module collects information on the resource utilization of VMs belonging to various PMs.
- SLA breaches are graded by analyzing if the resource utilization mechanism fits the criteria.
- The most energy-efficient devices may be identified through VM discovery.
- The relocation procedure has been completed in line with the PM, and if required, a new PM will be initiated. The method examines the current process node PMi to see if it is the Most Active PM. To gather background information on VMk resource usage, parameters such as PM List, VM List, PMi, f, rate, and an are specified.

The VM location algorithm avoids having the most active PM nodes averagely used, resulting in efficient workload balance and lower energy usage. Secure keys and cryptographic data are required for stable VM migration, and the TPM attached to the host provides a faith root that cannot be accessed from the original host platform. For primary and data migration strategies to work, stable protocols must be in place.

Results and Discussions

Cloud Sim is used to investigate the proposed EEARMM model, which is then compared to existing models like V-MAN and SCORCH. When VMs are created, an extra resource allocation assumption infers the real server's RAM. Each VM's memory factor varies dynamically and is derived using random functions. When the real machine's resource usage exceeds the mem SLA, which is more than the value of 'f,' the live virtual machine migration begins. The volume of virtual machine migrations at various rate thresholds was analyzed using various 'f' ranges. The suggested EEARMM consumes 20% less energy than existing models, illustrating how it saves money by using a more efficient virtual machine discovery method and an appropriate destination PM for dynamic migration. To guarantee that underutilized nodes are correctly assigned with particular workloads, EEARMM algorithms use optimum workload balancing. According to the graph, the number of computers that are least consumed has surpassed the number of machines that are most utilized. The use and management of resources is critical for properly managing workloads across computers(Maatoug et al., 2019; Mavromatis et al., 2020).

EEARMM is a dynamic resource allocation technique that considers process nodes in the Most Active, Minimum Active, and Inactive stages. It activates extra or new process nodes based on the determination, lowering energy consumption and prolonging the migration period. Cloud consolidation solutions strive to co-host virtual servers with numerous users on as few real machines as possible. This is accomplished through resource efficiency optimization, resource-aware scheduling to minimize connectivity costs, and resource-aware scheduling to reduce communication costs. The purpose of VM migration is to increase performance and exceed prior attempts, allowing administrators to keep the system functioning without hurting users. The energy consumption rate of the model is determined based on memory space utilization, resulting in fewer migrations owing to efficient system resource usage. Resource management and optimization are critical for efficient machine control. The efficacy of EEARMM in SLA enforcement is greatly enhanced over current ones, resulting in fewer SLA violations. The comparison of the V-MAN, SCORCH, and EARMM techniques delves into the important components of VM migration. The EARMM technique saves resources and removes SLA violations faster than the heuristics-based alternative. Alternative techniques, such as multidimensional optimization techniques, could be investigated.

EPADM

Energy prices make power usage an issue for server clusters. This chapter delves into high-performance power management approaches like as dynamic consolidation and dynamic power range use. Output isolation and virtualization overhead of several VMs are the main barriers for cloud consolidation. Cloud computing delivers utility-based IT services to worldwide clients, but data centers require a lot of space, power, and cash. A paradigm called Energy and Power-Aware Dynamic Migration

(EPADM) is developed to manage virtual resources in response to changing needs in order to decrease resource utilization and computation expenses.

The suggested EPADM model focuses on resource mapping and provisioning algorithms, virtual machine migration and consolidation, and eliminating SLA-based breaks. The model was tested using the CloudSim toolkit, and it outperformed others in terms of energy-power utilization and SLA violations(Pan et al., 2015; Talei et al., 2017). To categorize hosts and migrate virtual machines, researchers presented an energy-aware dynamic virtual machine migration method. Because static thresholds frequently fail to capture dynamic variations in cloud host load, this algorithm employs energy and power-aware dynamic migration to evaluate overload and under-loaded hosts. According to research, host load can be overloaded or underloaded.

Benefits

When designing a successful computational model, Cloud Service Providers (CSPs) must be adaptable and use efficient virtualization methods to ensure optimum energy and resource use. Cloud virtualization enables data centers to provide benefits such as resource utilization, scheduling, and routing.

Scalability: Customers expect a network that can accommodate multiple resource types.
Resource Management and Consolidation: Cloud virtual services can be combined with physical resources to achieve optimal energy utilisation.
Isolation: System output and vulnerabilities are isolated from each other.
Manageability: Integrate operations to effectively control resources.
System Robustness: Virtualization provides robustness to the cloud model.

Basic Procedure

Virtualization necessitates the consolidation of numerous virtual machines (VMs) that may handle varied user devices or specs. The services provided by the data center are separated into physical and virtual resource sets, with the virtual component serving as a management function for the physical resource collection. The Virtual Machine Monitor (VMM) is a piece of software that monitors virtual machines running on hardware. It consists of a collection of firmware, programs, and hardware that interface with the operating system and other external components. Physical services are asked to meet client demands based on SLA, pricing, and QoS standards. Live migration is a technique for moving a virtual machine to another PM dynamically based on the total amount of resource demands. It necessitates enhancements to the associated virtual machines and sophisticated organization. In sophisticated virtual machine administration, standard procedures are performed(Ahmad & Zhang, 2021; Marinakis et al., 2020; Rahmani et al., 2022).

VM Allocation: VM placement granted by VM development order to physical host.
VM Relocation: VM migration from PM when resource needs exceed physical host's resources.
VM Consolidation: Relocating VMs from active mode to inactive mode saves power and costs.

The approach focuses on two aspects of sophisticated VM management: energy use reduction and SLA violations. Increase the number of VMs per PM to address the difficulties, but this may result in resource inadequacy and/or SLA breaches. Furthermore, spreading virtual computers across a larger number of physical machines can increase energy consumption. To meet the aims of minimizing energy consumption and SLA breaches, EPADM is advised for balancing energy consumption and SLA breaches.

Proposed Model

EPADM is a dynamic approach for managing virtual machines (VMs) in data centers that identifies elements and situations to optimize Energy and Power Conscious Hybrid Migration. Following that, a protocol is proposed to achieve the following goals: decreasing energy use and minimizing SLA violations (Figure 3). When resources

Figure 3. Workload transfer by VM monitoring process

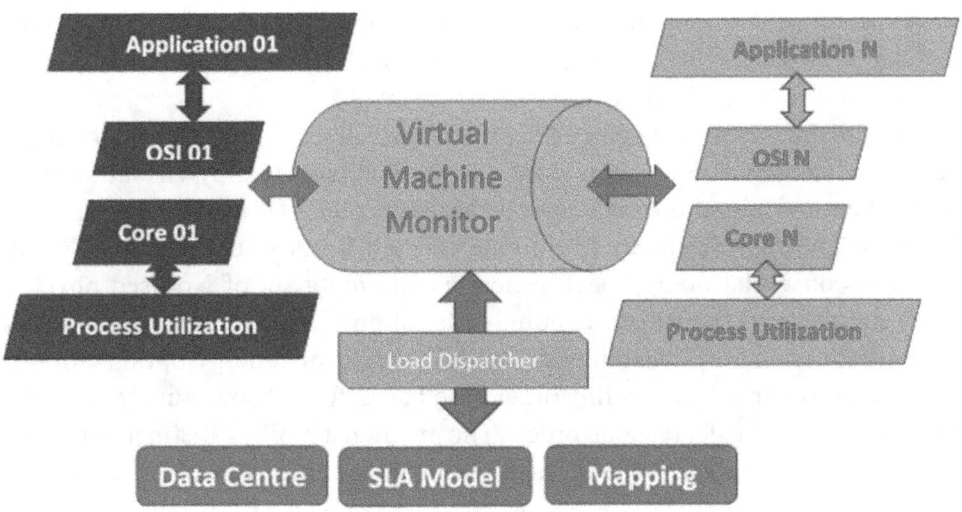

are unavailable, machine efficiency suffers as a result of the grade SLA breach. The total utilisation of the data center is the proportion of CPU power consumed. The CPU measure is assigned to each core based on its frequency, whereas the Power Adequacy 'PrAdph' is determined by the CPU measure per watt, denoted as CPUm/W. Power adequacy of a single physical host is calculated using the equation "PrAdph" to calculate operational use per utilized energy. When determining physical hosts for data center core hosts, optimal resource usage and maximum CPU power consumption are taken into account. Host consumption is the overall amount of CPU usage for all computers in active mode, generally known as the 'ON' condition. As the volume of host usage increases, the rate of energy consumption decreases. The suggested model specifies the power states of respective hosts as power ON, power hold, and power off. The host's power ON state is further classified into four categories depending on CPU utilization: fully utilized, somewhat utilized, little employed, and not utilized. With Max Utilized CPU and Min Utilized CPU, the host state can be modified to account for workload changes and complex VM migrations(Khajenasiri et al., 2017; Talei et al., 2017).

Purposes

The EPADM model seeks to reduce power consumption and SLA-V violations, and VM placement, relocation, and reorganization methods must be efficient to achieve these objectives. The VM placement mechanism distributes algorithmic stages onto physical hosts in the order of highest CPU consumption. The first-fit algorithm is used to select the target location, which can capacity check the physical host's ability to satisfy criteria. VM migration for energy savings classifies physical hosts and analyzes their CPU utilization, selecting the lowest workloads for the maximum utilisation. To guarantee that the CPU utilisation threshold surpasses max utilized CPU, the SLA-V reduction model runs a maximum utilised node check.

By shifting virtual machines from underutilized hosts to underutilized hosts, VM consolidation is applied to the burden generated by virtual machine deployment and migration procedures in the data center. This may be accomplished by moving virtual machines from underutilized hosts to underutilized hosts. The VM consolidation approach generates an empty set of No used physical cores and arranges them in descending order of power and CPU utilization. The SLA-V reduction VM consolidation model varies from energy optimization in that PHPt is ordered in ascending order based on CPU and power usage, whereas PHMn is listed in decreasing order. The method employs distinct threshold rates for energy and SLA-V reduction, with 95% for energy and 80% for SLA-V reduction(Boopathi, 2023; Boopathi, Siva Kumar, et al., 2023; Boopathi, Venkatesan, et al., 2023; Janardhana et al., 2023).

Dropping Energy Consumption

The EPADM paradigm allows for complicated policy juggling, evaluates data center variables, and analyzes SLA-V factors and energy efficiency to determine if active switching is required. The gap between desired and present energy efficiency is used to compute the energy efficiency ratio. When the factor value is directed to the target and the active mode's rate falls below the threshold rate, the switching mode in EPADM is engaged, allowing data centers to adapt to changing capital management requirements Because of the effective switching paradigm employed in data centers, the suggested EPADM achieves both energy savings and minimum SLA breaches.

Comparative Analysis

In Cloud, EPADM is introduced to evaluate its performance in a real-time cloud environment. The V-MAN, DEACM, and EEARMM models are used to correlate and assess the results. The model's inefficient dynamic resource management is evaluated using energy efficiency and SLA-V. The suggested EPADM model makes good use of the physical host by measuring and comparing its power consumption rate in order to achieve two goals: energy efficiency and SLA-V reduction. It consumes fewer resources than the other models under consideration, making it more cost-effective. When enforcing the SLA, the investigation discovered that the EPADM mode had the lowest SLA-V effects(Boopathi, Khare, et al., 2023; Mohanty et al., 2023; Palaniappan et al., 2023; Senthil et al., 2023; Vanitha et al., 2023). This was accomplished by continuously monitoring the number of virtual machine migrations. An energy-saving approach reduced migration while boosting physical host utilization, and effective VM migration strategies reduced SLA-based manipulations, resulting in restricted SLA-V. This aids in determining the EPADM model's success rate. The most significant concept is to utilize the model to determine the least used hosts and simulation time to guarantee that real hosts are not idle for a lengthy period of time, resulting in timely service to consumers. When simulation time is increased, the proposed model makes effective use of physical hosts through migration techniques, resulting in no underutilized hosts. The graph demonstrates that the suggested model's energy consumption rate is dictated by the CPU utilization factor, and that it performs fewer VM migrations than the compared ones by efficiently employing virtual cloud services.

FUTURE SCOPES

- By developing an optimization strategy with significant ideas, this research may be applied to real-time retail websites, financial cloud resource management, and educational applications. The review's findings are inconsistent with the observed outcomes.
- To find the contradiction, an algorithm must be created that compares it to another study.
- To analyse resource use, live VM migrations and resource utilization are required.

CONCLUSION

The EPADM prototype was developed using three algorithmic formats and various methodologies evaluated on the CloudSim cloud platform. The research aims, motivation, scope, background circumstances, methodology, and design illustration are covered. The Decentralized Energy-Aware Collaborative Model (DEACM) is proposed to provide QoS while focusing on energy and power-aware models. The DEACM allocates resources based on requirements using the Target selection method, consuming the least energy and minimizing VM migration. EEARMM, a framework that selects resources based on server monitoring and dynamic VM migration, uses fewer nodes and consumes less energy and simulation time than V-MAN and SCORCH.

EEARMM decreases memory utilization during VM migrations. The EPADM technique is assessed using physical host usage, power consumption, SLA breaches, and VM migrations. Existing research modules are also assessed to validate the technique.

- Cloud storage is a critical need for the globe, and significant organizations are investing in it. Space distribution, load balance, and energy saving are some of the research problems. Algorithms and strategies are being investigated to give solutions to these problems.
- For effective resource management of large-scale cloud data centers, a Shared Energy-Aware Collaboration Model (DEACM) was developed. It activated the VM selection mechanism for autonomous migration, prioritized proper workload balance, and allowed underutilized machines to sleep cycle. The two-level threshold technique for VM migration choices was implemented, and the simulation demonstrated improved workload balance while using less energy or being more efficient with resources.
- In this chapter, an EEARMM technique for complicated VM migration in a decentralized cloud system with collaborative performance is developed.

It recommends a VM discovery technique and a VM placement algorithm for migrating the selected VM to the correct destination host. The suggested study established its efficacy by changing workloads on a daily basis and including characteristics such as energy consumption, migration length, migration number, and SLA breaches.

- The EPADM model is used in the cloud to handle VMs dynamically, using algorithms aimed to accomplish energy and power-efficient data center management and minimise SLA breaches. The suggested EPADM model surpasses current simulations by offering better electricity pricing and power efficiency while resulting in fewer SLA violations. It is founded on three stages of effective cloud resource management.

REFERENCES

Ahmad, T., & Zhang, D. (2021). Using the internet of things in smart energy systems and networks. *Sustainable Cities and Society*, *68*, 102783. doi:10.1016/j.scs.2021.102783

Alhasnawi, B. N., Jasim, B. H., Esteban, M. D., & Guerrero, J. M. (2020). A novel smart energy management as a service over a cloud computing platform for nanogrid appliances. *Sustainability (Basel)*, *12*(22), 1–49. doi:10.3390u12229686

Alhasnawi, B. N., Jasim, B. H., Rahman, Z. A. S. A., & Siano, P. (2021). A novel robust smart energy management and demand reduction for smart homes based on internet of energy. *Sensors (Basel)*, *21*(14), 4756. doi:10.339021144756 PMID:34300497

Allahvirdizadeh, Y., Moghaddam, M. P., & Shayanfar, H. (2019). A survey on cloud computing in energy management of the smart grids. *International Transactions on Electrical Energy Systems*, *29*(10), e12094. doi:10.1002/2050-7038.12094

Anitha, C., R, K. C., Vivekanand, C. V., Lalitha, S. D., Boopathi, S., & R, R. (2023, February). Artificial Intelligence driven security model for Internet of Medical Things ({IoMT}). *IEEE Explore*. doi:10.1109/ICIPTM57143.2023.10117713

Babu, B. S., Kamalakannan, J., Meenatchi, N., M, S. K. S., S, K., & Boopathi, S. (2023). Economic impacts and reliability evaluation of battery by adopting Electric Vehicle. *IEEE Explore*, 1–6. doi:10.1109/ICPECTS56089.2022.10046786

Bera, S., Misra, S., & Rodrigues, J. J. P. C. (2015). Cloud Computing Applications for Smart Grid: A Survey. *IEEE Transactions on Parallel and Distributed Systems*, *26*(5), 1477–1494. doi:10.1109/TPDS.2014.2321378

Boopathi, S. (2019). Experimental investigation and parameter analysis of LPG refrigeration system using Taguchi method. *SN Applied Sciences*, *1*(8), 892. doi:10.100742452-019-0925-2

Boopathi, S. (2021). Improving of Green Sand-Mould Quality using Taguchi Technique. *Journal of Engineering Research*. doi:10.36909/jer.14079

Boopathi, S. (2022a). An Extensive Review on Sustainable Developments of Dry and Near-Dry Electrical Discharge Machining Processes. *Journal of Manufacturing Science and Engineering*, *144*(5), 50801. doi:10.1115/1.4052527

Boopathi, S. (2022b). Experimental investigation and multi-objective optimization of cryogenic Friction-stir-welding of AA2014 and AZ31B alloys using MOORA technique. *Materials Today. Communications*, *33*, 104937. doi:10.1016/j.mtcomm.2022.104937

Boopathi, S. (2022c). Performance Improvement of Eco-Friendly Near-Dry Wire-Cut Electrical Discharge Machining Process Using Coconut Oil-Mist Dielectric Fluid. *Journal of Advanced Manufacturing Systems*. doi:10.1142/S0219686723500178

Boopathi, S. (2023). An Investigation on Friction Stir Processing of Aluminum Alloy-Boron Carbide Surface Composite. In *Materials Horizons: From Nature to Nanomaterials* (pp. 249–257). Springer. doi:10.1007/978-981-19-7146-4_14

Boopathi, S., Arigela, S. H., Raman, R., Indhumathi, C., Kavitha, V., & Bhatt, B. C. (2023). Prominent Rule Control-based Internet of Things: Poultry Farm Management System. *IEEE Explore*, 1–6. doi:10.1109/ICPECTS56089.2022.10047039

Boopathi, S., Khare, R., Jaya Christiyan, K. G., Muni, T. V., & Khare, S. (2023). Additive manufacturing developments in the medical engineering field. In Development, Properties, and Industrial Applications of 3D Printed Polymer Composites (pp. 86–106). IGI Global. doi:10.4018/978-1-6684-6009-2.ch006

Boopathi, S., & Myilsamy, S. (2021). Material removal rate and surface roughness study on Near-dry wire electrical discharge Machining process. *Materials Today: Proceedings*, *45*(9), 8149–8156. doi:10.1016/j.matpr.2021.02.267

Boopathi, S., Siva Kumar, P. K., & Meena, R. S. J., S. I., P., S. K., & Sudhakar, M. (2023). Sustainable Developments of Modern Soil-Less Agro-Cultivation Systems. In Human Agro-Energy Optimization for Business and Industry (pp. 69–87). IGI Global. doi:10.4018/978-1-6684-4118-3.ch004

Boopathi, S., Thillaivanan, A., Pandian, M., Subbiah, R., & Shanmugam, P. (2021). Friction stir processing of boron carbide reinforced aluminium surface (Al-B4C) composite: Mechanical characteristics analysis. *Materials Today: Proceedings, 50*(5), 2430–2435. doi:10.1016/j.matpr.2021.10.261

Boopathi, S., Venkatesan, G., & Anton Savio Lewise, K. (2023). Mechanical Properties Analysis of Kenaf–Grewia–Hair Fiber-Reinforced Composite. In *Lecture Notes in Mechanical Engineering* (pp. 101–110). Springer. doi:10.1007/978-981-16-9057-0_11

Chang, Q., Chen, M., Ma, X., Gao, X., & Dehghani, M. (2021). A deep learning based secured energy management framework within a smart island. *Sustainable Cities and Society, 70*, 102938. doi:10.1016/j.scs.2021.102938

Domakonda, V. K., Farooq, S., Chinthamreddy, S., Puviarasi, R., Sudhakar, M., & Boopathi, S. (2023). Sustainable Developments of Hybrid Floating Solar Power Plants. In *Human Agro-Energy Optimization for Business and Industry* (pp. 148–167). IGI Global. doi:10.4018/978-1-6684-4118-3.ch008

Farid, F., Elkhodr, M., Sabrina, F., Ahamed, F., & Gide, E. (2021). A smart biometric identity management framework for personalised iot and cloud computing-based healthcare services. *Sensors (Basel), 21*(2), 1–18. doi:10.339021020552 PMID:33466730

Hashmi, S. A., Ali, C. F., & Zafar, S. (2021). Internet of things and cloud computing-based energy management system for demand side management in smart grid. *International Journal of Energy Research, 45*(1), 1007–1022. doi:10.1002/er.6141

Janardhana, K., Singh, V., Singh, S. N., Babu, T. S. R., Bano, S., & Boopathi, S. (2023). Utilization Process for Electronic Waste in Eco-Friendly Concrete: Experimental Study. In Sustainable Approaches and Strategies for E-Waste Management and Utilization (pp. 204–223). IGI Global.

Jeevanantham, Y. A., A, S., V, V., J, S. I., Boopathi, S., & Kumar, D. P. (2023). Implementation of Internet-of Things (IoT) in Soil Irrigation System. *IEEE Explore*, 1–5. doi:10.1109/ICPECTS56089.2022.10047185

Khajenasiri, I., Estebsari, A., Verhelst, M., & Gielen, G. (2017). A Review on Internet of Things Solutions for Intelligent Energy Control in Buildings for Smart City Applications. *Energy Procedia, 111*, 770–779. doi:10.1016/j.egypro.2017.03.239

Khayyam, H., Abawajy, J., Javadi, B., Goscinski, A., Stojcevski, A., & Bab-Hadiashar, A. (2013). Intelligent battery energy management and control for vehicle-to-grid via cloud computing network. *Applied Energy, 111*, 971–981. doi:10.1016/j.apenergy.2013.06.021

Kumara, V., Mohanaprakash, T. A., Fairooz, S., Jamal, K., Babu, T., & B., S. (2023). Experimental Study on a Reliable Smart Hydroponics System. In *Human Agro-Energy Optimization for Business and Industry* (pp. 27–45). IGI Global. doi:10.4018/978-1-6684-4118-3.ch002

Luján, E., Otero, A., Valenzuela, S., Mocskos, E., Steffenel, L. A., & Nesmachnow, S. (2019). Cloud Computing for Smart Energy Management (CC-SEM Project). *Communications in Computer and Information Science, 978*, 116–131. doi:10.1007/978-3-030-12804-3_10

Maatoug, A., Belalem, G., & Mahmoudi, S. (2019). Fog computing framework for location-based energy management in smart buildings. *Multiagent and Grid Systems, 15*(1), 39–56. doi:10.3233/MGS-190301

Marinakis, V., Doukas, H., Tsapelas, J., Mouzakitis, S., Sicilia, Á., Madrazo, L., & Sgouridis, S. (2020). From big data to smart energy services: An application for intelligent energy management. *Future Generation Computer Systems, 110*, 572–586. doi:10.1016/j.future.2018.04.062

Mavromatis, A., Colman-Meixner, C., Silva, A. P., Vasilakos, X., Nejabati, R., & Simeonidou, D. (2020). A Software-Defined IoT Device Management Framework for Edge and Cloud Computing. *IEEE Internet of Things Journal, 7*(3), 1718–1735. doi:10.1109/JIOT.2019.2949629

Mohanty, A., Venkateswaran, N., Ranjit, P. S., Tripathi, M. A., & Boopathi, S. (2023). Innovative Strategy for Profitable Automobile Industries: Working Capital Management. In Handbook of Research on Designing Sustainable Supply Chains to Achieve a Circular Economy (pp. 412–428). IGI Global.

Naveen, P., Ing, W. K., Danquah, M. K., Sidhu, A. S., & Abu-Siada, A. (2016). Cloud computing for energy management in smart grid - An application survey. *IOP Conference Series. Materials Science and Engineering, 121*(1), 12010. doi:10.1088/1757-899X/121/1/012010

Palaniappan, M., Tirlangi, S., Mohamed, M. J. S., Moorthy, R. M. S., Valeti, S. V., & Boopathi, S. (2023). Fused deposition modelling of polylactic acid (PLA)-based polymer composites: A case study. In Development, Properties, and Industrial Applications of 3D Printed Polymer Composites (pp. 66–85). IGI Global. doi:10.4018/978-1-6684-6009-2.ch005

Pan, J., Jain, R., Paul, S., Vu, T., Saifullah, A., & Sha, M. (2015). An Internet of Things Framework for Smart Energy in Buildings: Designs, Prototype, and Experiments. *IEEE Internet of Things Journal*, *2*(6), 527–537. doi:10.1109/JIOT.2015.2413397

Rahmani, M. K. I., Shuaib, M., Alam, S., Siddiqui, S. T., Ahmad, S., Bhatia, S., & Mashat, A. (2022). Blockchain-Based Trust Management Framework for Cloud Computing-Based Internet of Medical Things (IoMT): A Systematic Review. *Computational Intelligence and Neuroscience*, *2022*, 1–14. doi:10.1155/2022/9766844 PMID:35634070

Reddy, M. A., Reddy, B. M., Mukund, C. S., Venneti, K., Preethi, D. M. D., & Boopathi, S. (2023). Social Health Protection During the COVID-Pandemic Using IoT. In *The COVID-19 Pandemic and the Digitalization of Diplomacy* (pp. 204–235). IGI Global. doi:10.4018/978-1-7998-8394-4.ch009

Reka, S. S., & Ramesh, V. (2016). Demand side management scheme in smart grid with cloud computing approach using stochastic dynamic programming. *Perspectives in Science*, *8*, 169–171. doi:10.1016/j.pisc.2016.04.024

Sampath, B. C. S., & Myilsamy, S. (2022). Application of TOPSIS Optimization Technique in the Micro-Machining Process. In Trends, Paradigms, and Advances in Mechatronics Engineering (pp. 162–187). IGI Global. doi:10.4018/978-1-6684-5887-7.ch009

Selvakumar, S., Adithe, S., Isaac, J. S., Pradhan, R., Venkatesh, V., & Sampath, B. (2023). A Study of the Printed Circuit Board (PCB) E-Waste Recycling Process. In Sustainable Approaches and Strategies for E-Waste Management and Utilization (pp. 159–184). IGI Global.

Senthil, T. S. R. Ohmsakthi vel, Puviyarasan, M., Babu, S. R., Surakasi, R., & Sampath, B. (2023). Industrial Robot-Integrated Fused Deposition Modelling for the 3D Printing Process. In Development, Properties, and Industrial Applications of 3D Printed Polymer Composites (pp. 188–210). IGI Global. doi:10.4018/978-1-6684-6009-2.ch011

Sheikhi, A., Rayati, M., Bahrami, S., & Ranjbar, A. M. (2015a). Integrated demand side management game in smart energy hubs. *IEEE Transactions on Smart Grid*, *6*(2), 675–683. doi:10.1109/TSG.2014.2377020

Sheikhi, A., Rayati, M., Bahrami, S., & Ranjbar, A. M. (2015b). Integrated demand side management game in smart energy hubs. *IEEE Transactions on Smart Grid*, *6*(2), 675–683. doi:10.1109/TSG.2014.2377020

Sittón-Candanedo, I., Alonso, R. S., García, Ó., Muñoz, L., & Rodríguez-González, S. (2019). Edge computing, iot and social computing in smart energy scenarios. *Sensors (Basel), 19*(15), 3353. doi:10.339019153353 PMID:31370149

Subha, S., Inbamalar, T. M., R, K. C., Suresh, L. R., Boopathi, S., & Alaskar, K. (2023, February). A Remote Health Care Monitoring system using internet of medical things ({IoMT}). *IEEE Explore*. doi:10.1109/ICIPTM57143.2023.10118103

Talei, H., Essaaidi, M., & Benhaddou, D. (2017). Smart campus energy management system: Advantages, architectures, and the impact of using cloud computing. *ACM International Conference Proceeding Series, Part F1305*, (pp. 1–7). IEEE. 10.1145/3128128.3128129

Vanitha, S. K. R., & Boopathi, S. (2023). Artificial Intelligence Techniques in Water Purification and Utilization. In *Human Agro-Energy Optimization for Business and Industry* (pp. 202–218). IGI Global. doi:10.4018/978-1-6684-4118-3.ch010

Vennila, T., Karuna, M. S., Srivastava, B. K., Venugopal, J., Surakasi, R., & B., S. (2023). New Strategies in Treatment and Enzymatic Processes. In *Human Agro-Energy Optimization for Business and Industry* (pp. 219–240). IGI Global. doi:10.4018/978-1-6684-4118-3.ch011

Vignesh, S., Arulshri, K. P., Syedsajith, S., Kathiresan, S., Boopathi, S., & Dinesh Babu, P. (2018). Design and development of ornithopter and experimental analysis of flapping rate under various operating conditions. *Materials Today: Proceedings, 5*(11), 25185–25194. doi:10.1016/j.matpr.2018.10.320

Ye, J., Gao, Y., & Su, X. (2019). Smart Energy Management Cloud Platform Design Based on Offshore Wind Farm. *Proceedings - 2019 International Conference on Intelligent Transportation, Big Data and Smart City, ICITBS 2019*. IEEE, 10.1109/ICITBS.2019.00038

Yigit, M., Gungor, V. C., & Baktir, S. (2014). Cloud Computing for Smart Grid applications. *Computer Networks, 70*, 312–329. doi:10.1016/j.comnet.2014.06.007

Zhang, R., Cheng, X., & Yang, L. (2016). Energy Management Framework for Electric Vehicles in the Smart Grid: A Three-Party Game. *IEEE Communications Magazine, 54*(12), 93–101. doi:10.1109/MCOM.2016.1600335CM

Compilation of References

Abualigah, L. M. Q., & Hanandeh, E. S. (2015). Applying genetic algorithms to information retrieval using vector space model. *International Journal of Computer Science, Engineering and Applications*, *5*(1), 19. doi:10.5121/ijcsea.2015.5102

Afiq, H. (2023, April 13). Is Malaysia ready for commercial EVs? *The Malaysian Reserve*. https://themalaysianreserve.com/2023/04/13/is-malaysia-ready-for-commercial-evs/. https://themalaysianreserve.com/2023/04/13/is-malaysia-ready-for-commercial-evs/

Afonasova, M. A., Panfilova, E. E., Galichkina, M. A., & Ślusarczyk, B. (2019). Digitalization in economy and innovation: The effect on social and economic process. *Polish Journal of Management Studies*, *19*(2), 22–32. doi:10.17512/pjms.2019.19.2.02

Aggarwal, C. C., & Zhai, C. (2012). *Mining text data*. Springer Science & Business Media. doi:10.1007/978-1-4614-3223-4

Ahmad, T., & Zhang, D. (2021). Using the internet of things in smart energy systems and networks. *Sustainable Cities and Society*, *68*, 102783. doi:10.1016/j.scs.2021.102783

Akerman, A., Gaarder, I., & Mogstad, M. (2015). The Skill Complementarity of Broadband Internet. *The Quarterly Journal of Economics*, *30*(4), 1781–1824. doi:10.1093/qje/qjv028

Aksentijevic, N. K., Ježic, Z., & Zaninovic, P. A. (2021). The effects of information and communication technology (ICT) use on human development – A macroeconomic approach. *Economics*, *9*(128), 1–12.

Alam, M. S., Tanvir, M., Saha, D. K., & Das, S. K. (2021). Two Dimensional Convolutional Neural Network Approach for Real-Time Bangla Sign Language Characters Recognition and Translation. *SN Computer Science*, *2*(5), 1–13. doi:10.100742979-021-00783-6

Aleksandrova, A., Truntsevsky, Y., & Polutova, M. (2022). Digitalization and its impact on economic growth. *Revista de Economia Política*, *42*(2), 424–441. doi:10.1590/0101-31572022-3306

Alhasnawi, B. N., Jasim, B. H., Esteban, M. D., & Guerrero, J. M. (2020). A novel smart energy management as a service over a cloud computing platform for nanogrid appliances. *Sustainability (Basel)*, *12*(22), 1–49. doi:10.3390u12229686

Alhasnawi, B. N., Jasim, B. H., Rahman, Z. A. S. A., & Siano, P. (2021). A novel robust smart energy management and demand reduction for smart homes based on internet of energy. *Sensors (Basel)*, *21*(14), 4756. doi:10.339021144756 PMID:34300497

Ali, M. A., Alam, K., & Taylor, B. (2019). Incorporating affordability, efficiency, and quality in the ICT development index: Implications for index building and ICT policymaking. *The Information Society*, *36*(2), 71–96.

Allahvirdizadeh, Y., Moghaddam, M. P., & Shayanfar, H. (2019). A survey on cloud computing in energy management of the smart grids. *International Transactions on Electrical Energy Systems*, *29*(10), e12094. doi:10.1002/2050-7038.12094

Almunawar, M. N., Islam, Z., & Ordóñez de Pablos, P. (2022). *Digital Transformation Management: Challenges and Futures in the Asian Digital Economy*. Routledge. doi:10.4324/9781003224532

Al-Qahtani, M. F., Alsaffar, A. K., Alshammasi, A. R., Alsanni, G. S., Alyousef, Z. T., & Alhussaini, M. F. (2018). Social media in healthcare: advantages and challenges perceived by patients at a teaching hospital in eastern province. Saudi J Health Sciences

Alsughayr, A. R. (2015). Social media in healthcare: Uses, risks, and barriers. *Saudi Journal of Medicine & Medical Sciences*, *3*(2), 105. doi:10.4103/1658-631X.156405

Aly, H. (2020). Digital transformation, development and productivity in developing countries: Is artificial intelligence a cure or a blessing? *Review of Economics and Political Science*, *7*(4), 1–19.

Ambali, A. R., & Bakar, A. N. (2014). *Ict adoption and application in the Malaysian public sector*. Information Science Reference.

Aminur, R. A. (2020). M., Jasim, M., Haider ALI, M., & Hasanuzzaman, M. (n.d.). Bangla language modeling algorithm for automatic recognition of hand-sign-spelled Bangla sign language. *Frontiers of Computer Science*, *143302*(3), 143302. doi:10.100711704-018-7253-3

Amir, A. F., & Borhan, R. (2022). Students' perception and preference towards open and distance learning (Odl) during COVID-19 pandemic: A case study of studio-based learning. *Malaysian Journal of Sustainable Environment*, *9*(1), 285. doi:10.24191/myse.v9i1.17304

Andrés, L., & Cuberes, D. (2014). *The Diffusion of Internet : A Cross-Country Analysis*. Research Gate.

Angona, T. M., Shaon, A. S. M. S., Rashad Niloy, K. T., Karim, T., Tasnim, Z., Reza, S. M. S., & Mahbub, T. N. (2020). Automated bangla sign language translation system for alphabets by means of MobileNet. [Telecommunication Computing Electronics and Control]. *Telkomnika*, *18*(3), 1292–1301. doi:10.12928/telkomnika.v18i3.15311

Anitha, C., R, K. C., Vivekanand, C. V., Lalitha, S. D., Boopathi, S., & R, R. (2023, February). Artificial Intelligence driven security model for Internet of Medical Things ({IoMT}). *IEEE Explore*. doi:10.1109/ICIPTM57143.2023.10117713

Anshari, M., Almunawar, M. N., Lim, S. A., & Al-Mudimigh, A. (2019). Customer relationship management and big data enabled: Personalization & customization of services. *Applied Computing and Informatics*, *15*(2), 94–101. doi:10.1016/j.aci.2018.05.004

Antheunis, M. L., Tates, K., & Nieboer, T. E. (2013). Patients' and health professionals' use of social media in health care: Motives, barriers and expectations. *Patient Education and Counseling*, *92*(3), 426–431. doi:10.1016/j.pec.2013.06.020 PMID:23899831

APAC: active social media penetration by country 2022 . (n.d.). *Statista*. https://www.statista.com/statistics/255235/active-social-media-penetration-in-asian-countries/

Appiah-Otoo, I., & Song, N. (2021). The impact of ICT on economic growth-comparing rich and poor countries. *Telecommunications Policy*, *45*(2), 1–15. doi:10.1016/j.telpol.2020.102082

APR, A. P. R. (2022). *Prudential practice guide: draft CPG 229 climate change financial risks*. APR.

Arabi, A. M. K., & Allah, B. A. W. (2017). The impact of information and communication technology (ICT) development on economic growth in Sudan: An application of ARDL bounds testing approach. *Archives of Business Research*, *5*(3), 155–165.

Archambault, A., & Grudin, J. (2012). *A Longitudinal Study of Facebook, LinkedIn, & Twitter Use*. 2741–2750. ACM.

Ardichvili, A., Page, V., & Wentling, T. (2003). Motivation and barriers to participation in virtual knowledge-sharing communities of practice. *Journal of Knowledge Management*, *7*(1), 64–77. doi:10.1108/13673270310463626

Ashraf, S. N., & Singh, A. K. (2019). Does entrepreneurship ecosystem have a long-term relationship with economic growth in selected economies? A statistical investigation. In Misra S., Shukla S. and Batthini G. (Eds.), *Proceedings of the 13th Biennial Conference on Entrepreneurship, EDII Ahmedabad (Vol. 1*, pp. 176-187). Bookwell Publishing House, New Delhi.

Ashraf, S. N., & Singh, A. K. (2022). Implications of appropriate technology and farm inputs in the agricultural sector of Gujarat: Empirical analysis based on primary data. *Agricultural Economics Research Review*, *35*(2), 59–77. doi:10.5958/0974-0279.2022.00031.3

Atasoy, H. (2013). The effects of broadband internet expansion on labor market outcomes. *Industrial & Labor Relations Review*, *66*(2), 315–345. doi:10.1177/001979391306600202

Atwoli, L., Muhia, J., & Merali, Z. (2022). Mental health and climate change in Africa. *BJPsych International*, *19*(4), 86–89. doi:10.1192/bji.2022.14

Au, A. (2020). Reconceptualizing the generation in a digital(izing) modernity: Digital media, social networking sites, and the flattening of generations. *Journal for the Theory of Social Behaviour*, *50*(2), 163–183. doi:10.1111/jtsb.12239

Ayob, N. H., Aziz, M. A., & Ayob, N. A. (2022). Bridging the digital divide: Innovation policy and implementation in Malaysia. *International Journal of Academic Research in Business & Social Sciences*, *12*(8). doi:10.6007/IJARBSS/v12-i8/14554

Babu, B. S., Kamalakannan, J., Meenatchi, N., M, S. K. S., S, K., & Boopathi, S. (2023). Economic impacts and reliability evaluation of battery by adopting Electric Vehicle. *IEEE Explore*, 1–6. doi:10.1109/ICPECTS56089.2022.10046786

Bachmayer, W., & Klotz, J. (2021). *Homeoffice: Verbreitung*. Gestaltung, Meinungsbildung und Zukunft.

Bagozzi, R. P., & Dholakia, U. M. (2002). Intentional social action in virtual communities. *Journal of Interactive Marketing*, *16*(2), 2–21. doi:10.1002/dir.10006

Bahar, N., Wahab, S. N., & Ahmad, N. D. (2020). Understanding challenges faced in online teaching and learning among Malaysian universities' instructors during COVID-19 pandemic. *2020 Sixth International Conference on e-Learning (econf)*. IEEE. 10.1109/econf51404.2020.9385474

Bahruddin, U., & Ramadhanti F. S. (2020). Student's perceptions of Arabic online learning during COVID-19 emergency. *Journal for the Education of Gifted Young Scientists*, 1483-1492. doi:10.17478/jegys.763705

Baidoo-Anu, D. (2022). Digital divide in education: Exploring undergraduate Students' Online learning experiences in Ghana during the COVID-19 pandemic. *AERA*, *2022*. Advance online publication. doi:10.3102/IP.22.1885643

Bai, Y. (2017). The Faster, the Better? The Impact of Internet Speed on Employment. *Information Economics and Policy*, *40*, 21–25. doi:10.1016/j.infoecopol.2017.06.004

Bakari, S. (2022). The impact of digitalization and patent on economic growth in Romania. *MPRA Paper No. 114259*. MPRA. https://mpra.ub.uni-muenchen.de/114259/

Bakry, W., Nghiem, X. H., Farouk, S., & Vo, X. V. (2023). Does it hurt or help? revisiting the effects of ICT on economic growth and energy consumption: A nonlinear panel ARDL approach. *Economic Analysis and Policy*, *78*(1), 597–617. doi:10.1016/j.eap.2023.03.026

Balaji, M. S., & Roy, S. K. (2017). Value co-creation with Internet of things technology in the retail industry. *Journal of Marketing Management*, *33*(1–2), 7–31. doi:10.1080/0267257X.2016.1217914

Balsmeier, B., & Wörter, M. (2019). Is this time different? How digitalization influences job creation and destruction. *Research Policy*, *48*(8), 62–73. doi:10.1016/j.respol.2019.03.010

Bärenthaler-Sieber, S.; Bock-Schappelwein, J.; Böheim, M.; Kügler, A. and N. Schmidt-Padickakudy (2022). Digitalisation in Austria: Progress, Broadband Infrastructure and Open Access Networks. *WIFO Reports on Austria*. WIFO.

Barlett, Y. K., & Coulson, N. S. (2011). An investigation into the empowerment effects of using online support groups and how this affects health professional/patient communication. *Patient Education and Counseling*, *83*(1), 113–119. doi:10.1016/j.pec.2010.05.029 PMID:20599338

Basnin, N., Nahar, L., & Hossain, M. S. (2021). An integrated cnn-lstm model for bangla lexical sign language recognition. *Advances in Intelligent Systems and Computing*, *1309*, 695–707. doi:10.1007/978-981-33-4673-4_57

Basu, A., Walters, C., & Shepherd, M. (2003). Support vector machines for text categorization. *36th Annual Hawaii International Conference on System Sciences, 2003*. IEEE. 10.1109/HICSS.2003.1174243

Basu, S., & Fernald, J. (2007). Information and Communications Technology as a General-Purpose Technology: Evidence from US Industry Data. *German Economic Review*, *8*(2), 146–173. doi:10.1111/j.1468-0475.2007.00402.x

Benetoli, A., Chen, T. F., & Aslani, P. (2017). Consumer health-related activities on social media: Exploratory study. *Journal of Medical Internet Research*, *19*(10), e7656. doi:10.2196/jmir.7656 PMID:29030326

Bennett, K., Pitt, M., & Owers, M. (2010). Social Media Impact on Orginazations. In *5th Annual Conference Liverpool "BEAN."* Springer.

Bera, S., Misra, S., & Rodrigues, J. J. P. C. (2015). Cloud Computing Applications for Smart Grid: A Survey. *IEEE Transactions on Parallel and Distributed Systems*, *26*(5), 1477–1494. doi:10.1109/TPDS.2014.2321378

Bernama. (2020a). *Bernama: 12 Jan 2020: TM propelling smart tourism through 5G in Langkawi*. Official Portal of Ministry Of Communications And Digital. https://www.kkd.gov.my/en/public/news/16284-bernama-12-jan-2020-tm-propelling-smart-tourism-through-5g-in-langkawi

Bernama. (2020b). UMS student's girl in taking exam on a tree won viewers' hearts. *Astroawani*. https://www.astroawani.com/berita-malaysia/ums-students-grit-taking-exam-tree-won-viewers-hearts-247463

Bertschek, I. & Briglauer, W. (2018). *Wie die Digitale Transformation der Wirtschaft gelingt*. ZEW-Policy Brief 5/2018.

Bhaskaran, N., Kumar, M., & Janodia, M. D. (2017). Use of social media for seeking health related information–an exploratory study. *Journal of Young Pharmacists : JYP*, *9*(2), 267–271. doi:10.5530/jyp.2017.9.52

Bock-Schappelwein, J., Firgo, M., Kügler, A., & Schmidt-Padickakudy, N. (2021). *Digitalisierung in Österreich: Fortschritt, digitale Skills und Infrastrukturausstattung in Zeiten von Covid-19*. WIFO Monatsberichte 6/2021. https://www.wifo.ac.at/jart/prj3/wifo/resources/person_dokument/person_dokument.jart?publikationsid=67254&mime_type=application/pdfwifo.ac.at)

Bolton, R. N., Parasuraman, A., Hoefnagels, A., Kabadayi, S., Gruber, T., Loureiro, Y. K., Migchels, N., & Solnet, D. (2013). *Journal of Service Management, 24*(3), 245–267.

Bolton, R. N., McColl-Kennedy, J. R., Cheung, L., Gallan, A., Orsingher, C., Witell, L., & Zaki, M. (2018). Customer experience challenges: Bringing together digital, physical and social realms. *Journal of Service Management*, *29*(5), 776–808. doi:10.1108/JOSM-04-2018-0113

Bolton, R. N., Parasuraman, A., Hoefnagels, A., Migchels, N., Kabadayi, S., Gruber, T., Loureiro, Y. K., & Solnet, D. (2013). Understanding Generation Y and their use of social media: A review and research agenda. *Journal of Service Management*, *24*(3), 245–267. doi:10.1108/09564231311326987

Boopathi, S. (2021). Improving of Green Sand-Mould Quality using Taguchi Technique. *Journal of Engineering Research*. doi:10.36909/jer.14079

Boopathi, S., Arigela, S. H., Raman, R., Indhumathi, C., Kavitha, V., & Bhatt, B. C. (2023). Prominent Rule Control-based Internet of Things: Poultry Farm Management System. *IEEE Explore*, 1–6. doi:10.1109/ICPECTS56089.2022.10047039

Boopathi, S., Khare, R., Jaya Christiyan, K. G., Muni, T. V., & Khare, S. (2023). Additive manufacturing developments in the medical engineering field. In Development, Properties, and Industrial Applications of 3D Printed Polymer Composites (pp. 86–106). IGI Global. doi:10.4018/978-1-6684-6009-2.ch006

Boopathi, S., Siva Kumar, P. K., & Meena, R. S. J., S. I., P., S. K., & Sudhakar, M. (2023). Sustainable Developments of Modern Soil-Less Agro-Cultivation Systems. In Human Agro-Energy Optimization for Business and Industry (pp. 69–87). IGI Global. doi:10.4018/978-1-6684-4118-3.ch004

Boopathi, S. (2019). Experimental investigation and parameter analysis of LPG refrigeration system using Taguchi method. *SN Applied Sciences*, *1*(8), 892. doi:10.100742452-019-0925-2

Boopathi, S. (2022a). An Extensive Review on Sustainable Developments of Dry and Near-Dry Electrical Discharge Machining Processes. *Journal of Manufacturing Science and Engineering*, *144*(5), 50801. doi:10.1115/1.4052527

Boopathi, S. (2022b). Experimental investigation and multi-objective optimization of cryogenic Friction-stir-welding of AA2014 and AZ31B alloys using MOORA technique. *Materials Today. Communications*, *33*, 104937. doi:10.1016/j.mtcomm.2022.104937

Boopathi, S. (2022c). Performance Improvement of Eco-Friendly Near-Dry Wire-Cut Electrical Discharge Machining Process Using Coconut Oil-Mist Dielectric Fluid. *Journal of Advanced Manufacturing Systems*. doi:10.1142/S0219686723500178

Boopathi, S. (2023). An Investigation on Friction Stir Processing of Aluminum Alloy-Boron Carbide Surface Composite. In *Materials Horizons: From Nature to Nanomaterials* (pp. 249–257). Springer. doi:10.1007/978-981-19-7146-4_14

Boopathi, S., & Myilsamy, S. (2021). Material removal rate and surface roughness study on Near-dry wire electrical discharge Machining process. *Materials Today: Proceedings*, *45*(9), 8149–8156. doi:10.1016/j.matpr.2021.02.267

Boopathi, S., Thillaivanan, A., Pandian, M., Subbiah, R., & Shanmugam, P. (2021). Friction stir processing of boron carbide reinforced aluminium surface (Al-B4C) composite: Mechanical characteristics analysis. *Materials Today: Proceedings*, *50*(5), 2430–2435. doi:10.1016/j.matpr.2021.10.261

Boopathi, S., Venkatesan, G., & Anton Savio Lewise, K. (2023). Mechanical Properties Analysis of Kenaf–Grewia–Hair Fiber-Reinforced Composite. In *Lecture Notes in Mechanical Engineering* (pp. 101–110). Springer. doi:10.1007/978-981-16-9057-0_11

Bosslet, G. T., Torke, A. M., Hickman, S. E., Terry, C. L., & Helft, P. R. (2011). The patient–doctor relationship and online social networks: Results of a national survey. *Journal of General Internal Medicine*, *26*(10), 1168–1174. doi:10.100711606-011-1761-2 PMID:21706268

Bouaziz, S., Nabli, A., & Gargouri, F. (2019). Design a data warehouse schema from document-oriented database. *Procedia Computer Science*, *159*, 221–230. doi:10.1016/j.procs.2019.09.177

Boyd, D. & Ellison, N. B. (2008). Social Network Sites : Definition, History, and Scholarship. *Jcmc*, *13*, 210–230. doi:10.1111/j.1083-6101.2007.00393.x

Bresnahan, T., & Trajtenberg, M. (1995). General Purpose Technologies 'Engines of Growth'? *Journal of Econometrics*, *65*(1), 83–108. doi:10.1016/0304-4076(94)01598-T

Briglauer, W. and V. Stocker (2020). Bedeutung digitaler Infrastrukturen und Dienste und Maßnahmen zur Förderung der Resilienz in Krisenzeiten. *EcoAustria Policy Note 42.*

Briglauer, W., Cambini, C., & Grajek, M. (2018). Speeding Up the Internet: Regulation and Investment in the European Fiber Optic Infrastructure. *International Journal of Industrial Organization*, *61*(C), 613–652. doi:10.1016/j.ijindorg.2018.01.006

Briglauer, W., Dürr, N., & Gugler, K. (2021). A retrospective study on the regional benefits and spillover effects of high-speed broadband networks: Evidence from German counties. *International Journal of Industrial Organization*, *74*, 102677. doi:10.1016/j.ijindorg.2020.102677

Briglauer, W., & Gugler, K. (2019). Go for Gigabit? First Evidence on Economic Benefits of High-Speed Broadband Technologies in Europe. *Journal of Common Market Studies*, *57*(5), 1071–1090. doi:10.1111/jcms.12872

Briglauer, W., Whalley, J., & Stocker, V. (2020). *Public policy targets in EU broadband markets: The role of technology neutrality* (Vol. 44). Telecommunication Policy.

Broom, A. (2005). Virtually he@lthy: The impact of internet use on disease experience and the doctor-patient relationship. *Qualitative Health Research*, *15*(3), 325–345. doi:10.1177/1049732304272916 PMID:15761103

Brownlee, J. (2021). *Machine Learning Mastery With Python: Understand Your Data, Create Accurate*. Google Books. https://books.google.com.my/books?hl=en&lr=&id=BgmqDwAAQBAJ&oi=fnd&pg=PP1&dq=J+Brownlee+-+Machine+Learning+Mastery,+2016&ots=frp-TtkJ_O&sig=RLdyBCyOZ9EHKol45PLPV0uO0Ig&redir_esc=y#v=onepage&q&f=false

Brussevich, M., Dabla-Norris, E., & Khalid, S. (2022). Who bears the brunt of lockdown policies? Evidence from teleworkability measures across countries. *IMF Economic Review*, *70*(3), 560–589. doi:10.105741308-022-00165-9

Bulchand-Gidumal, J. (2020). Impact of Artificial Intelligence in Travel, Tourism, and Hospitality. Handbook of E-Tourism, 1–20. doi:10.1007/978-3-030-05324-6_110-1

Bunyan, J. (2020). Deputy minister who accused Sabah student Veveonah of lying says info came from Kudat MP. *Malay Mail.* https://www.malaymail.com/news/malaysia/2020/09/07/deputy-minister-who-accused-sabah-student-veveonah-of-lying-says-info-came/1900936

Buttcher, S., Clarke, C. L., & Cormack, G. V. (2016). *Information retrieval: Implementing and evaluating search engines.* Mit Press.

Caldwell, Z. (2020). Corporations and climate change: An investigation of mandatory climate risk disclosure in Australia. *Environmental and Planning Law, 37*(1), 3–17.

Calvo-Porral, C., & Pesqueira-Sanchez, R. (2020). Generational differences in technology behaviour: Comparing millennials and Generation X. *Kybernetes, 49*(11), 2755–2772. doi:10.1108/K-09-2019-0598

Campbell, L., Evans, Y., Pumper, M., & Moreno, M. A. (2016). Social media use by physicians: A qualitative study of the new frontier of medicine. *BMC Medical Informatics and Decision Making, 16*(1), 1–11. doi:10.118612911-016-0327-y PMID:27418201

Cao, B., Gupta, S., Wang, J., Hightow-Weidman, L. B., Muessig, K. E., Tang, W., Pan, S., Pendse, R., & Tucker, J. D. (2017). Social media interventions to promote HIV testing, linkage, adherence, and retention: Systematic review and meta-analysis. *Journal of Medical Internet Research, 19*(11), e7997. doi:10.2196/jmir.7997 PMID:29175811

Cao, X., Guo, X., Vogel, D., & Zhang, X. (2016). Exploring the influence of social media on employee work performance. *Internet Research, 26*(2), 529–545. doi:10.1108/IntR-11-2014-0299

Carbo-Valverde, S. (2017). The impact on digitalization on banking and financial stability. *Journal of Financial Management. Markets and Institutions, 5*(1), 133–140.

Castellacci, F., & Viñas-bardolet, C. (2019). *Internet use and job satisfaction.* Elesvier. doi:10.1016/j.chb.2018.09.001

Chang, Q., Chen, M., Ma, X., Gao, X., & Dehghani, M. (2021). A deep learning based secured energy management framework within a smart island. *Sustainable Cities and Society, 70*, 102938. doi:10.1016/j.scs.2021.102938

Charmaz, K. (2014). *Constructing grounded theory.* Sage.

Charoensukmongkol, P. (2014). Effects of support and job demands on social media use and work outcomes. *Computers in Human Behavior, 36*, 340–349. doi:10.1016/j.chb.2014.03.061

Chen, Y.C. (2017). Digital divide, digital inclusion, and digital opportunities. *Managing Digital Governance*, 54-75. doi:10.4324/9781315207667-10

Cheok, M. J., Omar, Z., & Jaward, M. H. (2019). A review of hand gesture and sign language recognition techniques. *International Journal of Machine Learning and Cybernetics, 10*(1), 131–153. doi:10.100713042-017-0705-5

Cheung, C. M. K., Chiu, P. Y., & Lee, M. K. O. (2011). Online social networks: Why do students use facebook? *Computers in Human Behavior, 27*(4), 1337–1343. doi:10.1016/j.chb.2010.07.028

Chipalkatti, N., Le, Q., & Rishi, M. (2021). Sustainability and society: Do environmental, social, and governance factors matter for foreign direct investment? *Sage, 14*(19), 6039.

Chung, H. L. (2023). Sarawak to launch blueprint to becoming leading digital economy by 2030. *The Edge Malaysia*. https://theedgemalaysia.com/article/sarawak-launch-blueprint-becoming-leading-digital-economy-2030

Clark, L., Pantidi, N., Cooney, O., Doyle, P., Garaialde, D., Edwards, J., Spillane, B., Gilmartin, E., Murad, C., & Munteanu, C. (2019). What makes a good conversation? Challenges in designing truly conversational agents. *Proceedings of the 2019 CHI Conference on Human Factors in Computing Systems*. ACM. 10.1145/3290605.3300705

Clerici, C. A., Veneroni, L., Bisogno, G., Trapuzzano, A., & Ferrari, A. (2012). Videos on rhabdomyosarcoma on YouTube: An example of the availability of information on pediatric tumors on the web. *Journal of Pediatric Hematology/Oncology, 34*(8), e329–e331. doi:10.1097/MPH.0b013e31825886f8 PMID:22858570

Crandall, R.W., Litan, R.E., Lehr, W. (2007). *The Effects of Broadband Deployment on Output and Employment: A Cross-Sectional Analysis of U.S. Data. Issues in Economic Policy.* The Brooklin Institution, Number 6.

Cristea, G., Vatamanescu, E.-M., & Mitan, A. (2017). *Managing People Dynamics Through the Lens of Generation Y.* 11Th International Management Conference: The Role of Management in the Economic Paradigm of the Xxist Century (Imc 2017), *Bucharest, Romania.*

Cui, G., Lockee, B., & Meng, C. (2013). Building modern online social presence: A review of social presence theory and its instructional design implications for future trends. *Education and Information Technologies, 18*(4), 661–685. doi:10.100710639-012-9192-1

Cutrer, C. M., & Pleil, A. M. (1991). Potential outcomes associated with direct-to-consumer advertising of prescription drugs: Physicians' perspectives. *Journal of Pharmaceutical Marketing & Management, 5*(3), 3–19.

Czernich, N. (2014). Does broadband internet reduce the unemployment rate? Evidence for Germany. *Information Economics and Policy, 29*, 32–45. doi:10.1016/j.infoecopol.2014.10.001

Czernich, N., Falck, O., Kretschmer, T., & Woessmann, L. (2011). Broadband Infrastructure and Economic Growth. *Economic Journal (London), 121*(552), 505–532. doi:10.1111/j.1468-0297.2011.02420.x

Dahlman, C., Mealy, S., & Wermelinger, M. (2016). *Harnessing the digital economy developing countries. OECD Development Centre,* (Working Paper No. 334). OECD Publishing, Paris.

Dahmani, M., Mabrouki, M., & Youssef, A. B. (2022). ICT, trade openness and economic growth in Tunisia: What is going wrong? *Economic Change and Restructuring, 55*(1), 2317–2336. doi:10.100710644-022-09388-2

Dawood, S. R. (2019). Digital divide and poverty eradication in the rural region of northern peninsular Malaysia. *The Indonesian Journal of Geography*, *51*(2), 172. doi:10.22146/ijg.37758

Dean, T. J., & Pacheco, D. F. (2014). Green marketing: A strategic balancing act for creating value. *The Journal of Business Strategy*, *35*(5), 14–22. doi:10.1108/JBS-11-2013-0109

Deb, K., Mony, H. P., & Chowdhury, S. (2012). Two-Handed Sign Language Recognition for Bangla Character Using Normalized Cross Correlation. *Global Journals.*

Desormeau, P. A., Walsh, K. M., & Segal, Z. V. (2018). Mindfulness-based stress reduction and mindfulness-based cognitive therapy. The Oxford Handbook of Stress and Mental Health, 688-704. Oxford Press. doi:10.1093/oxfordhb/9780190681777.013.33

Dijk, J. V. (2020). *The digital divide.* John Wiley & Sons.

Dixon, S. (2022). *Facebook users by country 2021.* Statista. https://www.statista.com/statistics/268136/top-15-countries-based-on-number-of-facebook-users/

Djafarova, E., & Rushworth, C. (2017). Exploring the credibility of online celebrities' Instagram profiles in influencing the purchase decisions of young female users. *Computers in Human Behavior*, *68*, 1–7. doi:10.1016/j.chb.2016.11.009

Dobrota, M., Jeremic, V., & Markovik, A. (2012). A new perspective of the ICT development index. *Information Development*, *28*(4), 271–280. doi:10.1177/0266666912446497

Doksil, M. (2020, June 24). UMS student who climbed hill, tree for better internet connectivity gets aid. *The Borneo Post.* https://www.theborneopost.com/2020/06/24/ums-student-who-climbed-hill-tree-for-better-internet-connectivity-gets-aid/

Domakonda, V. K., Farooq, S., Chinthamreddy, S., Puviarasi, R., Sudhakar, M., & Boopathi, S. (2023). Sustainable Developments of Hybrid Floating Solar Power Plants. In *Human Agro-Energy Optimization for Business and Industry* (pp. 148–167). IGI Global. doi:10.4018/978-1-6684-4118-3.ch008

Don, A., Zheleva, E., Gregory, M., Tarkan, S., Auvil, L., Clement, T., Shneiderman, B., & Plaisant, C. (2007). Discovering interesting usage patterns in text collections: integrating text mining with visualization. *Proceedings of the sixteenth ACM conference on Conference on information and knowledge management.* ACM. 10.1145/1321440.1321473

Dumais, S. T. (2004). Latent semantic analysis. *Annual Review of Information Science & Technology*, *38*(1), 188–230. doi:10.1002/aris.1440380105

Dutz, M. A., Orszag, J. M., & Willig, R. D. (2012). The liftoff of consumer benefits from the broadband revolution. *Review of Network Economics*, *11*(4). Advance online publication. doi:10.1515/1446-9022.1355

Economic Planning Unit. (2022). NATIONAL [Economic Planning Unit Prime Minister's Department.]. *Energy Policy*, 2022–2040.

Editoron. (2021, December 14). Budget 2022 to boost East Malaysia's development. *The Borneo Post*. https://www.theborneopost.com/2021/12/14/budget-2022-to-boost-East-malaysias-development/

Ellison, N. B., Gibbs, J. L., & Weber, M. S. (2014). The Use of Enterprise Social Network Sites for Knowledge Sharing in Distributed Organizations. *Sage*, *59*(1), 103–123. doi:10.1177/0002764214540510

Ellison, N. B., Steinfield, C., & Lampe, C. (2007). The Benefits of Facebook "' Friends :"' Social Capital and College Students '. *Journal of Computer-Mediated Communication*, *12*(4), 1143–1168. doi:10.1111/j.1083-6101.2007.00367.x

Erk, K., & Padó, S. (2008). A structured vector space model for word meaning in context. *Proceedings of the 2008 Conference on Empirical Methods in Natural Language Processing*. ACM. 10.3115/1613715.1613831

Erku, D. A., Belachew, S. A., Abrha, S., Sinnollareddy, M., Thomas, J., Steadman, K. J., & Tesfaye, W. H. (2021). When fear and misinformation go viral: Pharmacists' role in deterring medication misinformation during the'infodemic'surrounding COVID-19. *Research in Social & Administrative Pharmacy*, *17*(1), 1954–1963. doi:10.1016/j.sapharm.2020.04.032 PMID:32387230

European Commission. (2010). A Digital Agenda for Europe. Brussels. Available at: https://www.kowi.de/Portaldata/2/Resources/fp/2010-com-digital-agenda.pdf

European Commission. (2016). Connectivity for a Competitive Digital Single Market – Towards a European Gigabit Society. COM(2016)587 final, Brussels.

European Commission. (2021). *2030 Digital Compass: The European way for the Digital Decade*. Communication from the Commission to the European Parliament, the Council, the European Economic and Social Committee and the Committee of Regions.

European Commission. (2022a). *Digital Economy and Society Index (DESI) 2022*. Thematic Chapters.

European Commission. (2022b). Digital Economy and Society Index (DESI) 2022. Austria. Brussels.

European Commission. (2023). *Shaping Europe's digital future*. Europea. https://digital-strategy.ec.europa.eu/en

Eurostat. (2022). *Employed persons working from home as a percentage of the total employment*. Eurostat.

Fabling, R., & Grimes, A. (2021). Picking up speed: Does ultrafast broadband increase firm productivity? *Information Economics and Policy*, *57*, 100937. doi:10.1016/j.infoecopol.2021.100937

Face landmark detection guide . (n.d.). Google Developers. https://developers.google.com/mediapipe/solutions/vision/face_landmarker/

Fan, W., Wallace, L., Rich, S., & Zhang, Z. (2006). Tapping the power of text mining. *Communications of the ACM*, *49*(9), 76–82. doi:10.1145/1151030.1151032

Farhadi, M., Ismail, R., & Fooladi, M. (2012). Information and communication technology use and economic growth. *PLoS One*, *7*(11), e48903. doi:10.1371/journal.pone.0048903 PMID:23152817

Farid, F., Elkhodr, M., Sabrina, F., Ahamed, F., & Gide, E. (2021). A smart biometric identity management framework for personalised iot and cloud computing-based healthcare services. *Sensors (Basel)*, *21*(2), 1–18. doi:10.339021020552 PMID:33466730

Fatoki, O. (2019). Green Marketing Orientation and Environmental and Social Performance of Hospitality Firms in South Africa. *Foundations of Management*, *11*(1), 277–290. doi:10.2478/fman-2019-0023

Federal Ministry for Transport, Innovation and Technology (2019). *Breitbandstrategie 2030.* Österreichs Weg in die Gigabit-Gesellschaft. Vienna. breitbandstrategie2030_ua (1).pdf

Ferreira, D., Vale, M., Miguel Carmo, R., Encalada-Abarca, L., & Marcolin, C. (2021). The three levels of the urban digital divide: Bridging issues of coverage, usage and its outcomes in VGI platforms. *Geoforum*, *124*, 195–206. doi:10.1016/j.geoforum.2021.05.002

Fieseler, C., Meckel, M., & Ranzini, G. (2015). Professional Personae - How Organizational Identification Shapes Online Identity in the Workplace. *Journal of Computer-Mediated Communication*, *20*(2), 153–170. doi:10.1111/jcc4.12103

Fire, M., Goldschmidt, R., & Elovici, Y. (2014). Online social networks: Threats and solutions. *IEEE Communications Surveys and Tutorials*, *16*(4), 2019–2036. doi:10.1109/COMST.2014.2321628

Food and Drug Administration. (2020). *Generic Drugs: Questions and Answers*. FDA. https://www.fda.gov/drugs/questions-answers/generic-drugs-questions-answers

Ford, G. S. (2018). Is faster better? Quantifying the relationship between broadband speed and economic growth. *Telecommunications Policy*, *42*(9), 766–777. doi:10.1016/j.telpol.2018.05.006

Forman, C., Goldfarb, A., & Greenstein, S. (2012). The Internet and Local Wages: A Puzzle. *The American Economic Review*, *102*(1), 556–575. doi:10.1257/aer.102.1.556

Fox, S. (2011). *The social life of health information, 2011*. California Healthcare Foundation.

Fujita, S. (2021). *Income inequality in terms of a Gini coefficient: A Kaleckian perspective*.

Gartner, T. (2008). *Kernels for structured data*. World Scientific. doi:10.1142/6855

Gefen, D., & Straub, D. W. (2004). Consumer trust in B2C e-Commerce and the importance of social presence: Experiments in e-Products and e-Services. *Omega*, *32*(6), 407–424. doi:10.1016/j.omega.2004.01.006

Gerpott, T. J., & Ahmadi, N. (2015). Composite indices for the evaluation of a country's information technology development level: Extension of the IDI of the ITU. *Technological Forecasting and Social Change*, *98*(1), 174–185. doi:10.1016/j.techfore.2015.03.012

Giannakos, M. N., Chorianopoulos, K., Giotopoulos, K., & Vlamos, P. (2013). *Behaviour & Information Technology Using Facebook out of habit. October*. Taylor & Francis. doi:10.1080/0144929X.2012.659218

Gillet, S. E., Lehr, W. H., Osario, C. A., & Sirbu, M. (2006). Measuring the Economic Impact of Broadband Deployment. *Final Report. National Technical Assistance, Training, Research and Evaluation Project No. 99-07-13829.*

Gonzales, A. (2015). The contemporary US digital divide: From initial access to technology maintenance. *Information Communication and Society*, *19*(2), 234–248. doi:10.1080/1369118X.2015.1050438

Govender, J., & L. Govender, T. (2016). The influence of green marketing on consumer purchase behavior. *Environment and Ecology*, *7*(2), 77–85. doi:10.21511/ee.07(2).2016.8

Grajales, F. J. III, Sheps, S., Ho, K., Novak-Lauscher, H., & Eysenbach, G. (2014). Social media: A review and tutorial of applications in medicine and health care. *Journal of Medical Internet Research*, *16*(2), e2912. doi:10.2196/jmir.2912 PMID:24518354

Granovetter, M. (1983). The Strength of Weak Ties: A Network Theory Revisited. *Sociological Theory*, *1*, 201. doi:10.2307/202051

Greene, J. A., Choudhry, N. K., Kilabuk, E., & Shrank, W. H. (2011). Online social networking by patients with diabetes: A qualitative evaluation of communication with Facebook. *Journal of General Internal Medicine*, *26*(3), 287–292. doi:10.100711606-010-1526-3 PMID:20945113

Greenstein, S., & McDevitt, R. (2011). The Broadband Bonus: Estimating Broadband Internet's Economic Value. *Telecommunications Policy*, *35*(7), 617–632. doi:10.1016/j.telpol.2011.05.001

Gretzel, U., Sigala, M., Xiang, Z., & Koo, C. (2015). Smart tourism: Foundations and developments. *Electronic Markets*, *25*(3), 179–188. doi:10.100712525-015-0196-8

Grishman, R. (2019). Twenty-five years of information extraction. *Natural Language Engineering*, *25*(6), 677–692. doi:10.1017/S1351324919000512

Gruber, T. R. (1993). A translation approach to portable ontology specifications. *Knowledge Acquisition*, *5*(2), 199–220. doi:10.1006/knac.1993.1008

Guo, C., & Wan, B. (2022). The digital divide in online education in China during the COVID-19 pandemic. SSRN Electronic Journal. doi:10.2139/ssrn.4068301

Guo, J., Fan, Y., Pang, L., Yang, L., Ai, Q., Zamani, H., Wu, C., Croft, W. B., & Cheng, X. (2020). A deep look into neural ranking models for information retrieval. *Information Processing & Management*, *57*(6), 102067. doi:10.1016/j.ipm.2019.102067

Gupta, V., & Lehal, G. S. (2009). A survey of text mining techniques and applications. *Journal of Emerging Technologies in Web Intelligence*, *1*(1), 60–76. doi:10.4304/jetwi.1.1.60-76

Habibi, F., & Zabardast, M. A. (2020). Digitalization, education and economic growth: A comparative analysis of Middle East and OECD countries. *Technology in Society*, *63*(10), 23–47. doi:10.1016/j.techsoc.2020.101370

Haddud, A., Dugger, J. C., & Dugger, J. C. (2016). Social Media for Organizations. *Sage, 3*(1).

Haenssgen, M. J., Charoenboon, N., & Zanello, G. (2021). You've got a friend in me: How social networks and mobile phones facilitate healthcare access among marginalised groups in rural Thailand and Lao PDR. *World Development*, *137*, 105156. doi:10.1016/j.worlddev.2020.105156

Hall, C. M. (2011). Health and medical tourism: A kill or cure for global public health? *Tourism Review*, *66*(1/2), 4–15. doi:10.1108/16605371111127198

Han, X., Gao, T., Yao, Y., Ye, D., Liu, Z. & Sun, M. (2019). *OpenNRE: An open and extensible toolkit for neural relation extraction.*

Hanafizadeh, M. R., Hanafizadeh, P., & Saghaei, A. (2013). Perusing E-readiness and digital divide. *Digital Literacy*, 313-346. doi:10.4018/978-1-4666-1852-7.ch017

Hanafizadeh, M. R., Hanafizadeh, P., & Bohlin, E. (2013). Digital divide and E-Readiness. *International Journal of E-Adoption*, *5*(3), 30–75. doi:10.4018/ijea.2013070103

Haque, P., Das, B., & Kaspy, N. N. (2019, April 1). Two-Handed Bangla Sign Language Recognition Using Principal Component Analysis (PCA) and KNN Algorithm. *2nd International Conference on Electrical, Computer and Communication Engineering, ECCE 2019*. IEEE. 10.1109/ECACE.2019.8679185

Hargittai, E., & Hinnant, A. (2008). *Digital Inequality.*

Hasan, M., Sajib, T. H., & Dey, M. (2017, January 26). A machine learning based approach for the detection and recognition of Bangla sign language. *1st International Conference on Medical Engineering, Health Informatics and Technology, MediTec 2016*. IEEE. 10.1109/MEDITEC.2016.7835387

Hashmi, S. A., Ali, C. F., & Zafar, S. (2021). Internet of things and cloud computing-based energy management system for demand side management in smart grid. *International Journal of Energy Research*, *45*(1), 1007–1022. doi:10.1002/er.6141

Heimann, M., Bus, A., & Barr, R. (2021). Growing up in a digital world - Social and cognitive implications. Frontiers Media SA.

Hernandez, K., Faith, B., Martín, P. P., & Ramalingam, B. (2016). *The impact of digital technology on economic growth and productivity, and its implications for employment and equality: An evidence review. Evidence Report No. 207*. Institute of Development Studies., https://www.ids.ac.uk/publications/the-impact-of-digital-technology-on-economic-growth-and-productivity-and-its-implications-for-employment-and-equality-an-evidence-review/

Holt, L., & Jamison, M. (2009). Broadband and contributions to economic growth: Lessons from the US experience. *Telecommunications Policy*, *33*(10-11), 575–581. doi:10.1016/j.telpol.2009.08.008

Honnibal, M. & Montani, I. (2017). spaCy 2: Natural language understanding with Bloom embeddings, convolutional neural networks and incremental parsing. *Sage, 7*(1), 411-420.

Horng, S. M., & Wu, C. L. (2020). How behaviors on social network sites and online social capital influence social commerce intentions. *Information & Management, 57*(2), 103176. doi:10.1016/j.im.2019.103176

Househ, M., Borycki, E., & Kushniruk, A. (2014). Empowering patients through social media: The benefits and challenges. *Health Informatics Journal, 20*(1), 50–58. doi:10.1177/1460458213476969 PMID:24550564

Huub, R., Bondarouk, T., & Looise, K. J. (2023). E-HRM: Innovation or Irritation. An Explorative Empirical Study in Five Large Companies on Web-based HRM on JSTOR. *Management Review, 15*(3), 364–380. https://www.jstor.org/stable/41783479

Hwang, G. J., & Chien, S. Y. (2022). Definition, roles, and potential research issues of the metaverse in education: An artificial intelligence perspective. In *Computers and Education* (Vol. 3). Artificial Intelligence., doi:10.1016/j.caeai.2022.100082

Iftikhar, R., & Abaalkhail, B. (2017). Health-seeking influence reflected by online health-related messages received on social media: Cross-sectional survey. *Journal of Medical Internet Research, 19*(11), e5989. doi:10.2196/jmir.5989 PMID:29146568

Immordino-yang, M. H., Christodoulou, J. A., & Singh, V. (2012). *Perspectives on Psychological Science Rest Is Not Idleness : Implications of.* Sage. doi:10.1177/1745691612447308

India - social media users 2023. (2020). Statista.www.statista.com › statistics › number-of-social-network-users-in-india.

India: COVID-19 impact on social media app usage 2020 . (n.d.). Statista. www.statista.com. Retrieved August 1, 2022, from https://www.statista.com/statistics/1114459/india-coronavirus-impact-on-weekly-usage-time-of-social-networking-apps/

India: social media penetration 2022 . (n.d.). Statista. https://www.statista.com/statistics/284436/india-social-media-penetration/

Irtyshcheva, I., Stehnei, M., Popadynets, N., Bogatyrev, K., Boiko, Y., Kramarenko, I., Senkevich, O., Hryshyna, N., Kozak, I., & Ishchenko, O. (2021). The effect of digital technology development on economic growth. *International Journal of Data and Network Science, 5*(1), 25–36. doi:10.5267/j.ijdns.2020.11.006

Islalm, M. S., Rahman, M. M., Rahman, M. H., Arifuzzaman, M., Sassi, R., & Aktaruzzaman, M. (2019, September 1). Recognition Bangla Sign Language using Convolutional Neural Network. *2019 International Conference on Innovation and Intelligence for Informatics, Computing, and Technologies (3ICT).* IEEE. 10.1109/3ICT.2019.8910301

Islam, M. M., Uddin, M. R., Ferdous, M. J., Akter, S., & Nasim Akhtar, M. (2022). BdSLW-11: Dataset of Bangladeshi sign language words for recognizing 11 daily useful BdSL words. *Data in Brief, 45*, 108747. doi:10.1016/j.dib.2022.108747 PMID:36425983

Islam, M. S., Mousumi, S. S. S., Azad Rabby, A. K. M. S., Hossain, S. A., & Abujar, S. (2018). A Potent Model to Recognize Bangla Sign Language Digits Using Convolutional Neural Network. *Procedia Computer Science*, *143*, 611–618. doi:10.1016/j.procs.2018.10.438

Ivus, O., & Boland, M. (2015). The employment and wage impact of broadband deployment in Canada. *The Canadian Journal of Economics. Revue Canadienne d'Economique*, *48*(5), 1803–1830. doi:10.1111/caje.12180

Jacobs, J., & Subramaniam, P. (2020, October 23). Cover story: The digital divide and disconnection. *The Edge Markets.* https://www.theedgemarkets.com/article/cover-story-digital-divide-and-disconnection

Jaeger, L., & Eckhardt, A. (2021). Eyes wide open: The role of situational information security awareness for security-related behaviour. *Information Systems Journal*, *31*(3), 429–472. doi:10.1111/isj.12317

Jamil, S. (2021). From digital divide to digital inclusion: Challenges for wide-ranging digitalization in Pakistan. *Telecommunications Policy*, *45*(8), 102206. doi:10.1016/j.telpol.2021.102206

Janardhana, K., Singh, V., Singh, S. N., Babu, T. S. R., Bano, S., & Boopathi, S. (2023). Utilization Process for Electronic Waste in Eco-Friendly Concrete: Experimental Study. In Sustainable Approaches and Strategies for E-Waste Management and Utilization (pp. 204–223). IGI Global.

Janz, N. K., & Becker, M. H. (1984). The health belief model: A decade later. *Health Education Quarterly*, *11*(1), 1–47. doi:10.1177/109019818401100101 PMID:6392204

Jeevanantham, Y. A., A, S., V, V., J, S. I., Boopathi, S., & Kumar, D. P. (2023). Implementation of Internet-of Things (IoT) in Soil Irrigation System. *IEEE Explore*, 1–5. doi:10.1109/ICPECTS56089.2022.10047185

Jio, S., & Sun, Q. (2021). Digital economic development and its impact on economic growth in China: Research based on the perspective of sustainability. *Sustainability (Basel)*, *13*(1), 1–18. doi:10.3390u131810245

Jo, T. (2019). Text mining. *Studies in Big Data.*

Johari, Z. (2020, July 19). Grouses over poor internet access. *The Boreno Post.* https://www.theborneopost.com/2020/07/19/grouses-over-poor-internet-access/

Jovanovic, B., & Rousseau, P. (2005). General Purpose Technologies. In Ph. Aghion & St. Durlauf (Eds.), *Handbook of Economic Growth* (Vol. 1, pp. 1181–1224). Part B.

Jyoti, B., & Singh, A. K. (2023). Does digitalization have a causal relationship with economic development? An experience from a country-wise panel data statistical analysis. In R. S. Nazlı & G. Sari (Eds.), *Perspectives on Society and Technology Addiction*. IGI Global. doi:10.4018/978-1-6684-8397-8.ch003

Kadhim, A. I. (2019). Survey on supervised machine learning techniques for automatic text classification. *Artificial Intelligence Review*, *52*(1), 273–292. doi:10.100710462-018-09677-1

Kannan, S., V. Gurusamy, S. Vijayarani, J. Ilamathi, M. Nithya, S. Kannan, V. J. I. J. o. C. S. Gurusamy and C. Networks (2014). Preprocessing techniques for text mining. *Sage, 5*(1), 7-16.

Katoch, S., Singh, V., & Tiwary, U. S. (2022). Indian Sign Language recognition system using SURF with SVM and CNN. *Array (New York, N.Y.), 14*, 100141. doi:10.1016/j.array.2022.100141

Katz, R., & Callorda, F. (2020). *Assessing the Economic Potential of 10 G Networks*. Telecom Advisory Services.

Kementerian Komunikasi Dan Multimedia Malaysia. (2018). *Communications & multimedia blueprint 2018-2025 digitise and humanise* (1st ed.). Ministry of Communications and Multimedia.

Khajenasiri, I., Estebsari, A., Verhelst, M., & Gielen, G. (2017). A Review on Internet of Things Solutions for Intelligent Energy Control in Buildings for Smart City Applications. *Energy Procedia, 111*, 770–779. doi:10.1016/j.egypro.2017.03.239

Khal, K., Bilgihan, A., Okumus, F., & Cobanoglu, C. (2013). Generation Y travelers ' commitment to online social network websites. *Tourism Management, 35*, 13–22. doi:10.1016/j.tourman.2012.05.005

Khanagha, S., Volberda, H., & Oshri, I. (2017). Customer Co-Creation and Exploration of Emerging Technologies: The Mediating Role of Managerial Attention and Initiatives. *Long Range Planning, 50*(2), 221–242. doi:10.1016/j.lrp.2015.12.019

Khayyam, H., Abawajy, J., Javadi, B., Goscinski, A., Stojcevski, A., & Bab-Hadiashar, A. (2013). Intelligent battery energy management and control for vehicle-to-grid via cloud computing network. *Applied Energy, 111*, 971–981. doi:10.1016/j.apenergy.2013.06.021

Kim, H. Lee, J., & Oh, S. E. (2019). Individual characteristics influencing the sharing of knowledge on social networking services: online identity, self-efficacy, and knowledge sharing intentions. *Sage, 39*(4), 379–390. doi:10.1080/0144929X.2019.1598494

Kim, H., & Scott, C. (2019). Change communication and the use of anonymous social media at work: Implications for employee engagement. *Corporate Communications, 24*(3), 410–424. doi:10.1108/CCIJ-07-2018-0076

Kim, W., Jeong, O. R., & Lee, S. W. (2010). On social Web sites. *Information Systems, 35*(2), 215–236. doi:10.1016/j.is.2009.08.003

Kiron, D., Palmer, D., Phillips, A. N., & Kruschwitz, N. (2012). *Social Business: What Are Companies Really Doing?* Sage.

Klaus, M. (2017). *Digital and the real world, the: Computational foundations of mathematics, science, technology, and philosophy*. World Scientific.

Kolko, J. (2012). Broadband and Local Growth. *Journal of Urban Economics, 71*(1), 100–113. doi:10.1016/j.jue.2011.07.004

Korhonen, A., Ó Séaghdha, D., Silins, I., Sun, L., Högberg, J., & Stenius, U. (2012). Text mining for literature review and knowledge discovery in cancer risk assessment and research. *PLoS One*, *7*(4), e33427. doi:10.1371/journal.pone.0033427 PMID:22511921

Koutroumpis, P. (2009). The Economic Impact of Broadband on Growth: A Simultaneous Approach. *Telecommunications Policy*, *33*(9), 471–485. doi:10.1016/j.telpol.2009.07.004

Kowsari, K., Brown, D. E., Heidarysafa, M., Meimandi, K. J., Gerber, M. S., & Barnes, L. E. (2017). Hdltex: Hierarchical deep learning for text classification. *2017 16th IEEE international conference on machine learning and applications (ICMLA)*. IEEE. 10.1109/ICMLA.2017.0-134

Kreitem, H. M. (2020). Weaponization of access, communication inequalities as a form of control: Case of Israel/Palestine. *Digital Inequalities in the Global South*, 137-157. doi:10.1007/978-3-030-32706-4_7

Kreutzer, T. (2009). *Generation Mobile : Online and Digital Media Usage on Mobile Phones among Low-Income Urban Youth in South Africa. February*. Research Gate.

Kumar, A., & Sharma, P. (2013). Impact of climate variation on agricultural productivity and food security in rural India. *Economics Discussion Papers, No. 2013-43*. Kiel Institute for the World Economy Germany. http://www.economics-ejournal.org/economics/discussionpapers/2013-43

Kumara, V., Mohanaprakash, T. A., Fairooz, S., Jamal, K., Babu, T., & B., S. (2023). Experimental Study on a Reliable Smart Hydroponics System. In *Human Agro-Energy Optimization for Business and Industry* (pp. 27–45). IGI Global. doi:10.4018/978-1-6684-4118-3.ch002

Kumar, A., Ahmad, M. M., & Sharma, P. (2015). Carbon emission and global food security: A cross country analysis. *PENCIL Publication of Agricultural Sciences*, *2*(1), 7–24.

Kumar, A., Ahmad, M. M., & Sharma, P. (2017). Influence of climatic and non-climatic factors on sustainable food security in India: A statistical investigation. *International Journal of Sustainable Agricultural Management and Informatics*, *3*(1), 1–30. doi:10.1504/IJSAMI.2017.082917

Kumar, P. (2015). Green marketing innovations in small Indian firms. *World Journal of Entrepreneurship, Management and Sustainable Development*, *11*(3), 176–190. doi:10.1108/WJEMSD-01-2015-0003

Kumar, P. (2016). State of green marketing research over 25 years (1990-2014): Literature survey and classification. *Marketing Intelligence & Planning*, *34*(1), 137–158. doi:10.1108/MIP-03-2015-0061

Kumpu, V. (2022). What is public engagement and how does it help to address climate change? A review of climate communication research. *Environmental Communication*, *16*(3), 304–316. doi:10.1080/17524032.2022.2055601

Kupperschmidt, B. (2000). *Multigeneration Employees: Strategies for Effective Management*. NIH.

Kuráková, I., Vallušová, A., & Marasová, J. (2021). Measuring the digital divide in the V4 countries using the digital divide index. *Ekonomika a spoločnosť, 22*(1), 77-93. doi:10.24040/eas.2021.22.1.77-93

Kutsuri, G. N., Kamberdieva, S. S., Dedegkaev, V. K. H., Sopoeva, I. A., & Shelkunova, T. G. (2019). Impact of digitalization on improvement of economy, IT and internet of business. *Journal of Physics: Conference Series, 1399*(3), 1–6. doi:10.1088/1742-6596/1399/3/033008

Kwon, M.-W., D'Angelo, J., & McLeod, D. M. (2013). Facebook Use and Social Capital. *Sage, 33*(1–2), 35–43. doi:10.1177/0270467613496767

Kyytsönen, M., Ikonen, J., Aalto, A. M., & Vehko, T. (2022). The self-assessed information security skills of the Finnish population: A regression analysis. *Computers & Security, 118*, 102732. doi:10.1016/j.cose.2022.102732

Lahtinen, O., & Salmivalli, C. (2020). An effectiveness study of a digital mindfulness-based program for upper secondary education students. *Mindfulness, 11*(11), 2494–2505. doi:10.100712671-020-01462-y

Lai, K. H., Wu, S. J., & Wong, C. W. Y. (2013). Did reverse logistics practices hit the triple bottom line of Chinese manufacturers? *International Journal of Production Economics, 146*(1), 106–117. doi:10.1016/j.ijpe.2013.03.005

Lal, M. (2015). *Neo4j graph data modeling*. Packt Publishing Ltd.

Lampe, C., Ellison, N., & Steinfield, C. (2007). *A Familiar Face (book): Profile Elements as Signals in an Online Social Network*. 435–444.

Latha, K., Meena, K. S., Pravitha, M. R., Dasgupta, M., & Chaturvedi, S. K. (2020). Effective use of social media platforms for promotion of mental health awareness. *Journal of Education and Health Promotion, 9*(1), 9. doi:10.4103/jehp.jehp_90_20 PMID:32642480

Laurent, P., Chollet, T., & Herzberg, E. (2015). *Intelligent automation entering the business world.* Deloitte. https://www2.deloitte.com/content/dam/Deloitte/lu/Documents/operatihttps://www2.deloitte.com/content/dam/Deloitte/no/Documents/technology/intelligent-automation-entering-the-business-world.pdf

Lee, Y. C., & Wu, W. L. (2014). The effects of situated learning and health knowledge involvement on health communications. *Reproductive Health, 11*(1), 1–7. doi:10.1186/1742-4755-11-93 PMID:25542070

Lefebvre, R. C., & Bornkessel, A. S. (2013). Digital social networks and health. *Circulation, 127*(17), 1829–1836. doi:10.1161/CIRCULATIONAHA.112.000897 PMID:23630086

Leonardi, P. M., & Treem, J. W. (2012). Knowledge management technology as a stage for strategic self-presentation: Implications for knowledge sharing in organizations. *Information and Organization, 22*(1), 37–59. doi:10.1016/j.infoandorg.2011.10.003

Levac, J. J., & O'Sullivan, T. (2010). Social media and its use in health promotion. Revue interdisciplinaire des sciences de la santé-Interdisciplinary. *Journal of Health Science*, *1*, 47–53.

Lewins, A., & Silver, C. (2007). Qualitative coding in software: Principles and processes. *Using Software in Qualitative Research.*, *9780857025012*.

Lewis, H. (2005). Defining product stewardship and sustainability in the Australian packaging industry. *Environmental Science & Policy*, *8*(1), 45–55. doi:10.1016/j.envsci.2004.09.002

Li, R., & Gospodarik, C. G. (2021). The impact of digital economy on economic growth based on pearson correlation test analysis. In: J. Jansen, B., Liang, H., Ye, J. (eds) *International Conference on Cognitive based Information Processing and Applications (CIPA 2021). Lecture Notes on Data Engineering and Communications Technologies, vol 85.* Springer, Singapore. 10.1007/978-981-16-5854-9_3

Liao, T. (2015). Augmented or admented reality? The influence of marketing on augmented reality technologies. *Information Communication and Society*, *18*(3), 310–326. doi:10.1080/1369118X.2014.989252

Líšková, Z. D., Cudlínová, E., Pártlová, P., & Petr, D. (2016). Importance of Green Marketing and Its Potential. *Visegrad Journal on Bioeconomy and Sustainable Development*, *5*(2), 61–64. doi:10.1515/vjbsd-2016-0012

Li, Y., & Wang, X. (2018). Seeking health information on social media: A perspective of trust, self-determination, and social support. [JOEUC]. *Journal of Organizational and End User Computing*, *30*(1), 1–22. doi:10.4018/JOEUC.2018010101

Loane, S. S., & D'Alessandro, S. (2014). Empowered and knowledgeable health consumers: The impact of online support groups on the doctor–patient relationship. [AMJ]. *Australasian Marketing Journal*, *22*(3), 238–245. doi:10.1016/j.ausmj.2014.08.007

Lowenthal, P. R. (2020). *Social Presence*. IGI Global. doi:10.4018/978-1-60566-984-7.ch011

lu, L., Bock, D., & Joseph, M. (2013). Green marketing: What the Millennials buy. *Journal of Business Strategy, 34*(6), 3–10. doi:10.1108/JBS-05-2013-0036

Luccioni, A., & Palacios, H. (2019). Using natural language processing to analyze financial climate disclosures. *Proceedings of the 36th International Conference on Machine Learning*, Long Beach, California.

Luján, E., Otero, A., Valenzuela, S., Mocskos, E., Steffenel, L. A., & Nesmachnow, S. (2019). Cloud Computing for Smart Energy Management (CC-SEM Project). *Communications in Computer and Information Science*, *978*, 116–131. doi:10.1007/978-3-030-12804-3_10

Maatoug, A., Belalem, G., & Mahmoudi, S. (2019). Fog computing framework for location-based energy management in smart buildings. *Multiagent and Grid Systems*, *15*(1), 39–56. doi:10.3233/MGS-190301

Macey, W. H. (2009). *Employee engagement : tools for analysis, practice, and competitive advantage*. Sage.

Mackey, T. K., Cuomo, R. E., Guerra, C., & Liang, B. A. (2015). The rise of digital direct-to-consumer advertising?: Comparison of direct-to-consumer advertising expenditure trends from publicly available data sources and global policy implications. *BMC Health Services Research, 15*(1), 236. doi:10.118612913-015-0885-1 PMID:26084705

Ma, D., & Zhu, Q. (2022). Innovation on emerging economies: Research on the digital economy driving high quality green development. *Journal of Business Research, 145*(1), 891–813. doi:10.1016/j.jbusres.2022.03.041

Magsamen-Conrad, K., Upadhyaya, S., Joa, C. Y., & Dowd, J. (2015). Bridging the divide: Using UTAUT to predict multigenerational tablet adoption practices. *Computers in Human Behavior, 50*, 186–196. doi:10.1016/j.chb.2015.03.032 PMID:25937699

Mahmoud, A. B., Fuxman, L., Mohr, I., Reisel, W. D., & Grigoriou, N. (2021). "We aren't your reincarnation!" workplace motivation across X, Y and Z generations. *International Journal of Manpower, 42*(1), 193–209. doi:10.1108/IJM-09-2019-0448

Malay Mail. (2023, March 17). *Sabah state govt completes three digital hubs*. Malay Mail. https://www.malaymail.com/news/malaysia/2023/03/17/sabah-state-govt-completes-three-digital-hubs/60050

Malaysian communications and multimedia commission. (2023). *Capacity Development Programme (CDeP) 2023 The Evolution of The Mobile Cellular Communication: From 0G to 5G*. https://www.mcmc.gov.my/en/capacity-development-programme-cdep/capacity-development-programme-cdep-2023

Malik, M., Tahir, M. J., Jabbar, R., Ahmed, A., & Hussain, R. (2020). Self-medication during Covid-19 pandemic: Challenges and opportunities. *Drugs & Therapy Perspectives, 36*(12), 565–567. doi:10.100740267-020-00785-z PMID:33041621

Mannai, M., Karâa, W. B. A., & Ghezala, H. H. B. (2018). Information extraction approaches: A survey. Information and Communication Technology. Springer.

Marinakis, V., Doukas, H., Tsapelas, J., Mouzakitis, S., Sicilia, Á., Madrazo, L., & Sgouridis, S. (2020). From big data to smart energy services: An application for intelligent energy management. *Future Generation Computer Systems, 110*, 572–586. doi:10.1016/j.future.2018.04.062

Markham, M. J., Gentile, D., & Graham, D. L. (2017). Social media for networking, professional development, and patient engagement. *American Society of Clinical Oncology Educational Book, 37*(37), 782–787. doi:10.1200/EDBK_180077 PMID:28561727

Martínez, P. (2015). Customer loyalty: Exploring its antecedents from a green marketing perspective. *International Journal of Contemporary Hospitality Management, 27*(5), 896–917. doi:10.1108/IJCHM-03-2014-0115

Massari, L. (2009). Analysis of MySpace user profiles. *Information Systems Frontiers 2009 12*(4), 361–367. doi:10.1007/s10796-009-9206-8

Mavromatis, A., Colman-Meixner, C., Silva, A. P., Vasilakos, X., Nejabati, R., & Simeonidou, D. (2020). A Software-Defined IoT Device Management Framework for Edge and Cloud Computing. *IEEE Internet of Things Journal, 7*(3), 1718–1735. doi:10.1109/JIOT.2019.2949629

Ma, W. W. K., & Chan, A. (2014). Knowledge sharing and social media: Altruism, perceived online attachment motivation, and perceived online relationship commitment. *Computers in Human Behavior, 39*, 51–58. doi:10.1016/j.chb.2014.06.015

Mazzocchini, F. J., & Lucarelli, C. (2022). Success or failure in equity crowdfunding? A systematic literature review and research perspectives. *Management Research Review*. doi:10.1108/MRR-09-2021-0672

McGilvray, J. (2014). Chomsky: Language, mind and politics. *Polity.*

Meister, J. C., & Willyerd, K. (2010). *Mentoring Millennials: Spotlight on Leadership -- The Next Generation*. HBR. www.hbr.org

Melville, N., Kraemer, K., & Gurbaxani, V. (2004). Review: Information technology and organizational performance: An integrative model of it business value. *Management Information Systems Quarterly, 28*(2), 283–322. doi:10.2307/25148636

MIGHT. (2014). *Malaysian Rail Supporting Industry Roadmap 2030*. Malaysian Industry-Government Group for High Technology.

Minaee, S., Kalchbrenner, N., Cambria, E., Nikzad, N., Chenaghlu, M., & Gao, J. (2021). Deep Learning—based Text Classification: A Comprehensive Review. *ACM Computing Surveys, 54*(3), 1–40. doi:10.1145/3439726

Minges, M. (2015). *Exploring the relationship between broadband and economic growth.* Background Paper, World Development Report 2016.

Minister of Finance. (2022). *Budget Speech 2023*. MOF. chrome-extension://efaidnbmnnnibpcajpcglclefindmkaj/https://budget.mof.gov.my/pdf/2023/ucapan/buku-budget-speech-2023.pdf

Mitra, B., & Craswell, N. (2017). Neural models for information retrieval. arXiv preprint arXiv:1705.01509.

Moens, M.-F. (2006). *Information extraction: algorithms and prospects in a retrieval context.* Springer Science & Business Media.

Mohanty, A., Venkateswaran, N., Ranjit, P. S., Tripathi, M. A., & Boopathi, S. (2023). Innovative Strategy for Profitable Automobile Industries: Working Capital Management. In Handbook of Research on Designing Sustainable Supply Chains to Achieve a Circular Economy (pp. 412–428). IGI Global.

Moorhead, S. A., Hazlett, D. E., Harrison, L., Carroll, J. K., Irwin, A., & Hoving, C. (2013). A new dimension of health care: Systematic review of the uses, benefits, and limitations of social media for health communication. *Journal of Medical Internet Research*, *15*(4), e1933. doi:10.2196/jmir.1933 PMID:23615206

Moser, S. C. (2016). Reflections on climate change communication research and practice in the second decade of the 21st century: What more is there to say? *Wiley Interdisciplinary Reviews: Climate Change*, *7*(3), 345–369. doi:10.1002/wcc.403

Muhammad, A. (2022, April 8). 'I travel 10-15km daily just to get internet': Telupid student attends online class in car boot. *Daily express online*. https://www.dailyexpress.com.my/news/190389/i-travel-10-15km-daily-just-to-get-internet-telupid-student-attends-online-class-in-car-boot-/

Mukhopadhyay, N., & Sengupta, P. P. (2021). *Gini inequality index: Methods and applications*. CRC Press. doi:10.1201/9781003143642

Myovella, G., Karacuka, M., & Haucap, J. (2020). Digitalization and economic growth: A comparative analysis of Sub-Saharan Africa and OECD economies. *Telecommunication Policy, 44*(2).

Myrick, J. G., Holton, A. E., Himelboim, I., & Love, B. (2016). # Stupidcancer: Exploring a typology of social support and the role of emotional expression in a social media community. *Health Communication*, *31*(5), 596–605. doi:10.1080/10410236.2014.981664 PMID:26453480

Naim, M. F., & Lenka, U. (2017). Linking knowledge sharing, competency development, and affective commitment: Evidence from Indian Gen Y employees. *Journal of Knowledge Management*, *21*(4), 885–906. doi:10.1108/JKM-08-2016-0334

Narula, S. A., & Desore, A. (2016). Framing green consumer behaviour research: Opportunities and challenges. *Social Responsibility Journal*, *12*(1), 1–22. doi:10.1108/SRJ-08-2014-0112

National Water Research Institute. (2021). *Climate change adaptation framework for water sectors*. National water research institute of malaysia ministry of environment and water.

Naveen, P., Ing, W. K., Danquah, M. K., Sidhu, A. S., & Abu-Siada, A. (2016). Cloud computing for energy management in smart grid - An application survey. *IOP Conference Series. Materials Science and Engineering*, *121*(1), 12010. doi:10.1088/1757-899X/121/1/012010

Ngai, E. W. T., Moon, K. L. K., Lam, S. S., Chin, E. S. K., & Tao, S. S. C. (2015). Social media models, technologies, and applications: An academic review and case study. *Industrial Management & Data Systems*, *115*(5), 769–802. doi:10.1108/IMDS-03-2015-0075

Nguyen, V. B. (2021). The digitalization – economic growth relationship in developing countries and the role of governance. *Scientific Annals of Economics and Business*, *68*(4), 481–493. doi:10.47743aeb-2021-0028

Nifakos, S., Chandramouli, K., Nikolaou, C. K., Papachristou, P., Koch, S., Panaousis, E., & Bonacina, S. (2021). Influence of human factors on cyber security within healthcare organisations: A systematic review. *Sensors (Basel)*, *21*(15), 5119. doi:10.339021155119 PMID:34372354

Nihal, R. A., Broti, N. M., Deowan, S. A., & Rahman, S. (2021). Design and Development of a Humanoid Robot for Sign Language Interpretation. *SN Computer Science*, *2*(3), 1–17. doi:10.100742979-021-00627-3

Nistoreanu, P., Aluculesei, A. C., & Avram, D. (2020). Is green marketing a label for ecotourism? The Romanian experience. *Information (Basel)*, *11*(8), 1–13. doi:10.3390/info11080389

Nkwe, N., & Cohen, J. (2017). *Impact Of Social Network Sites On Psychological And Behavioural Outcomes In The Work-Place.*

Nordin, M., Grenestam, E., & Gullstrand, J. (2019). Is Super-fast Broadband Negative? An IV-Estimation of the Broadband Effect on Firms' Sales and Employment Level. School of Economics and Management. *Working Paper*, *2019*(8).

OECD. (2001). Understanding the Digital Divide. Digital Economy Papers. No. 49. Paris.

OECD. (2001). *Understanding the digital divide*. OECD Digital Economy Papers. doi:10.1787/20716826

OECD. (2018). Bridging the Rural Digital Divide. OECD Digital Economy Papers. No. 265. Paris.

Ogheneovo, E., & Japheth, R. (2016). Application of vector space model to query ranking and information retrieval. *International Journal of Advanced Research in Computer Science and Software Engineering*, *6*(5).

Ogloblina, E. V., Seredina, M. I., Altunina, J. O., Kodolov, V. A., & Lebedev, K. A. (2020). Socio-economic consequences of digital development of the economy. *Revista Inclusiones*, *7*(1), 421–430.

Oh, H. J., Lauckner, C., Boehmer, J., Fewins-Bliss, R., & Li, K. (2013). Facebooking for health: An examination into the solicitation and effects of health-related social support on social networking sites. *Computers in Human Behavior*, *29*(5), 2072–2080. doi:10.1016/j.chb.2013.04.017

Oh, H. J., & Lee, B. (2012). The effect of computer-mediated social support in online communities on patient empowerment and doctor–patient communication. *Health Communication*, *27*(1), 30–41. doi:10.1080/10410236.2011.567449 PMID:21797714

Oksa, R., Pirkkalainen, H., Salo, M., Savela, N., & Oksanen, A. (2022). Professional social media-enabled productivity: A five-wave longitudinal study on the role of professional social media invasion, work engagement and work exhaustion. *Information Technology & People*, *35*(8), 349–368. doi:10.1108/ITP-11-2021-0899

On-device, Real-time Body Pose Tracking with MediaPipe BlazePose – Google AI Blog. (n.d.). Retrieved May 15, 2023, from https://ai.googleblog.com/2020/08/on-device-real-time-body-pose-tracking.html

One, T. M. (2023, February 8). *Pushing Malaysia's smart city development in 2022*. TMONE. https://www.tmone.com.my/resources/think-tank/article/smart-city-development-in-2022/

Ordóñez de Pablos, P., Zhang, X., Almunawar, M. N., & Labra Gayo, J. E. (2022). *Handbook of Research on Big Data, Green Growth, and Technology Disruption in Asian Companies and Societies*. IGI-Global. doi:10.4018/978-1-7998-8524-5

Ordóñez de Pablos, P., Zhang, X., & Almunawar, N. (2022b). *Green, Circular, and Digital Economies as Tools for Recovery and Sustainability*. IGI-Global.

Özsoy, S., Ergüzel, O. Ş., Ersoy, A. Y., & Saygılı, M. (2022). The impact of digitalization on export of high technology products: A panel data approach. *The Journal of International Trade & Economic Development, 31*(2), 277–298. doi:10.1080/09638199.2021.1965645

Palaniappan, M., Tirlangi, S., Mohamed, M. J. S., Moorthy, R. M. S., Valeti, S. V., & Boopathi, S. (2023). Fused deposition modelling of polylactic acid (PLA)-based polymer composites: A case study. In Development, Properties, and Industrial Applications of 3D Printed Polymer Composites (pp. 66–85). IGI Global. doi:10.4018/978-1-6684-6009-2.ch005

Pan, J., Jain, R., Paul, S., Vu, T., Saifullah, A., & Sha, M. (2015). An Internet of Things Framework for Smart Energy in Buildings: Designs, Prototype, and Experiments. *IEEE Internet of Things Journal, 2*(6), 527–537. doi:10.1109/JIOT.2015.2413397

Papageorgiou, A., Strigkos, M., Politou, E., Alepis, E., Solanas, A., & Patsakis, C. (2018). Security and privacy analysis of mobile health applications: The alarming state of practice. *IEEE Access : Practical Innovations, Open Solutions, 6*, 9390–9403. doi:10.1109/ACCESS.2018.2799522

Paroutis, S., & Al Saleh, A. (2009). Determinants of knowledge sharing using Web 2.0 technologies. *Journal of Knowledge Management, 13*(4), 52–63. doi:10.1108/13673270910971824

Pearce, W., Grundmann, R., Hulme, M., Raman, S., Hadley Kershaw, E., & Tsouvalis, J. (2017). Beyond counting climate consensus. *Environmental Communication, 11*(6), 723–730. doi:10.1080/17524032.2017.1333965

Peattie, K. (2023). *Towards sustainability : The third age of green marketing*. Research Gate.

Piaget, J. (1964). Part I: Cognitive development in children: Piaget development and learning. *Journal of Research in Science Teaching, 2*(3), 176–186. doi:10.1002/tea.3660020306

Pimmer, C., Linxen, S., & Gröhbiel, U. (2012). Facebook as a learning tool? A case study on the appropriation of social network sites from mobile phones in developing countries. *British Journal of Educational Technology, 43*(5), 726–738. doi:10.1111/j.1467-8535.2012.01351.x

Podder, K. K., Chowdhury, M. E. H., Tahir, A. M., Mahbub, Z., Khandakar, A., Hossain, M. S., & Kadir, M. A. (2022). Bangla Sign Language (BdSL) Alphabets and Numerals Classification Using a Deep Learning Model. *Sensors (Basel), 22*(2), 574. Advance online publication. doi:10.339022020574 PMID:35062533

Polonsky, M. J. (1994). An introduction to green marketing. *Electronic Green Journal, 1*(2). doi:10.5070/G31210177

Porter, M. F. (1980). *An algorithm for suffix stripping*. Program. doi:10.1108/eb046814

Pow, H. (2010). Doctors caught revealing secret patient information in Facebook posts. *Sunday Telegraph, 5.*

Pradhan, R. P., Sarangi, A. K., Maity, C., & Behera, R. R. (2022). ICT infrastructure and economic growth in G-20 Countries: New insights form ARDL modelling. *Journal of Developing Areas, 56*(1), 47–58. doi:10.1353/jda.2022.0009

Pribeanu, C., & Balog, A. (n.d.). *The 12 th International Scientific Conference eLearning and Software for Education Bucharest, Explaining The Acceptance Of Facebook By University Students With The Uses & Gratifications Theory*. eLSE. 10.12753/2066-026X-15-013

Prime Minister Malaysia. (2015). *Eleventh Malaysia plan, 2016-2020: Anchoring growth on people*. Unit Perancang Ekonomi.

Prochaska, J. J., Coughlin, S. S., & Lyons, E. J. (2017). Social media and mobile technology for cancer prevention and treatment. *American Society of Clinical Oncology Educational Book, 37*(37), 128–137. doi:10.1200/EDBK_173841 PMID:28561647

Pujara, J., & Singh, S. (2018). Mining knowledge graphs from text. *Proceedings of the Eleventh ACM International Conference on Web Search and Data Mining*. ACM. 10.1145/3159652.3162011

Qiang, C. Z.-W., & Rossotto, C. M. (2009). Economic impacts of broadband. In *Information and Communications for Development: Extending Reach and Increasing Impact*. World Bank, Global Information and Communication Technologies Department.

Quan-haase, A., & Young, A. L. (2010). *Bulletin of Science, Technology & Society Uses and Gratifications of of Facebook and Instant Messaging*. Sage. doi:10.1177/0270467610380009

Raacke, J., Ph, D., Bonds-raacke, J., & Ph, D. (2008). *MySpace and Facebook : Applying the Uses and Gratifications Theory to Exploring*. Mary Ann Liebert Pub. doi:10.1089/cpb.2007.0056

Raeskyesa, D. G. S., & Lukas, E. N. (2019). Does digitalization increase economic growth? Evidence from ASEAN8 countries. *Jurnal Ekonomi Indonesia, 8*(2), 267–278. doi:10.52813/jei.v8i2.33

Rafi, A. M., Nawal, N., Bayev, N. S. N., Nima, L., Shahnaz, C., & Fattah, S. A. (2019, October 1). Image-based Bengali Sign Language Alphabet Recognition for Deaf and Dumb Community. *2019 IEEE Global Humanitarian Technology Conference, GHTC 2019*. IEEE. 10.1109/GHTC46095.2019.9033031

Ragnedda, M., & Muschert, G. W. (2013). *The Digital Divide. The Internet and Social Inequality in International Perspective*. Routledge. doi:10.4324/9780203069769

Rahmani, M. K. I., Shuaib, M., Alam, S., Siddiqui, S. T., Ahmad, S., Bhatia, S., & Mashat, A. (2022). Blockchain-Based Trust Management Framework for Cloud Computing-Based Internet of Medical Things (IoMT): A Systematic Review. *Computational Intelligence and Neuroscience, 2022*, 1–14. doi:10.1155/2022/9766844 PMID:35634070

Rath, B. N., & Hermawan, D. (2019). Do information and communication technologies foster economic growth in Indonesia? *Bulletin of Monetary Economics and Banking*, *22*(1), 103–122. doi:10.21098/bemp.v22i1.1041

Reddy, M. A., Reddy, B. M., Mukund, C. S., Venneti, K., Preethi, D. M. D., & Boopathi, S. (2023). Social Health Protection During the COVID-Pandemic Using IoT. In *The COVID-19 Pandemic and the Digitalization of Diplomacy* (pp. 204–235). IGI Global. doi:10.4018/978-1-7998-8394-4.ch009

Reddy, R. (2006). in Support of Society. *IEEE Intelligent Systems*, *21*(3).

Reibel, D., & McCown, D. (2019). Mindfulness-based stress reduction. Handbook of Mindfulness-Based Programmes, 29-41. Oxford Press. doi:10.4324/9781315265438-4

Reka, S. S., & Ramesh, V. (2016). Demand side management scheme in smart grid with cloud computing approach using stochastic dynamic programming. *Perspectives in Science*, *8*, 169–171. doi:10.1016/j.pisc.2016.04.024

Rolnick, D., Donti, P., Kaack, L., Kochanski, K., Lacoste, A., Sankaran, K., Ross, A., Milojevic-Dupont, N., Jaques, N., & Waldman-Brown, A. (2022). Tackling climate change with machine learning. *Sage, 55*(2), 1-96.

Rosenstock, I. M. (1974). Historical origins of the health belief model. *Health Education Monographs*, *2*(4), 328–335. doi:10.1177/109019817400200403

Roshdi, A., & Roohparvar, A. (2015). Information retrieval techniques and applications. *International Journal of Computer Networks and Communications Security*, *3*(9), 373–377.

Rossmann, A., Ranjan, K. R., & Sugathan, P. (2016). Drivers of user engagement in eWoM communication Drivers of user engagement in eWoM communication (Journal of Services Marketing). *Journal of Services Marketing*, *C*(January), 541–553. doi:10.1108/JSM-01-2015-0013

Rouhani, S., Ashrafi, A., Zare Ravasan, A., & Afshari, S. (2016). The impact model of business intelligence on decision support and organizational benefits. *Journal of Enterprise Information Management*, *29*(1), 19–50. doi:10.1108/JEIM-12-2014-0126

RTR (2022). *RTR Internet Monitor.* 2. Quartal 2022. lLOLnMDg5uSnEx67-report-q2-22-internet-monitor.pdf (23degrees.io)

Rybak, N., & Hassall, M. (2021). Deep Learning Unsupervised Text-Based Detection of Anomalies in US Chemical Safety and Hazard Investigation Board Reports. *2021 International Conference on Electrical, Computer, Communications and Mechatronics Engineering (ICECCME)*. IEEE. 10.1109/ICECCME52200.2021.9590834

Sabban, I., Lopez, C., & Mercuzot, Y. (2020). Automatic analysis of insurance reports through deep neural networks to identify severe claims. *Annals of Actuarial Science*, 1-26.

Sadun, R. E., & Schanberg, L. E. (2018). Using social media to promote medication adherence. *Nature Reviews. Rheumatology*, *14*(8), 445–446. doi:10.103841584-018-0037-1 PMID:29915383

Sahoo, D., Nayak, S., & Behera, J. (2021). *Digitalization and Economic Performance of Two Fast-Growing Asian Economies: India and the People's Republic of China. ADBI Working Paper 1243*. Asian Development Bank Institute. Available https://www.adb.org/publications/digitalization-economic-performance-fast-growing-asian-economies-india-prc

Saif, A., & Mahayuddin, Z. R. (2021). An efficient method for hand gesture recognition using robust features vector Authors AFM Saif, Zainal Rasyid Mahayuddin. *Journal Information System and Technology Management (JISTM)*. https://scholar.google.com/citations?view_op=view_citation&hl=en&user=1Y50UjsAAAAJ&sortby=pubdate&citation_for_view=1Y50UjsAAAAJ:hFOr9nPyWt4C

Saif, A., & Mahayuddin, Z. R. (2023). View artiAugmented Reality-Based 3D Human Hands Tracking From Monocular True Images Using Convolutional Neural Networkcle. *Handbook of Research on Artificial Intelligence and Knowledge Management in Asia's Digital Economy*. IGI Global. https://scholar.google.com/citations?view_op=view_citation&hl=en&user=1Y50UjsAAAAJ&sortby=pubdate&citation_for_view=1Y50UjsAAAAJ:e5wmG9Sq2KIC

Salton, G., Wong, A., & Yang, C.-S. (1975). A vector space model for automatic indexing. *Communications of the ACM*, *18*(11), 613–620. doi:10.1145/361219.361220

Samala, N., Katkam, B. S., Bellamkonda, R. S., & Rodriguez, R. V. (2022). Impact of AI and robotics in the tourism sector: A critical insight. *Journal of Tourism Futures*, *8*(1), 73–87. doi:10.1108/JTF-07-2019-0065

Sampath, B. C. S., & Myilsamy, S. (2022). Application of TOPSIS Optimization Technique in the Micro-Machining Process. In Trends, Paradigms, and Advances in Mechatronics Engineering (pp. 162–187). IGI Global. doi:10.4018/978-1-6684-5887-7.ch009

Sanderson, H. (2019). How do climate risks affect corporations and how could they address these risks? *Sage, 1,* 1-6.

Sanzidul Islam, M., Sultana Sharmin Mousumi, S., Jessan, N. A., Shahariar Azad Rabby, A., & Akhter Hossain, S. (2018, November 30). Ishara-Lipi: The First Complete MultipurposeOpen Access Dataset of Isolated Characters for Bangla Sign Language. *2018 International Conference on Bangla Speech and Language Processing, ICBSLP 2018*. IEEE. 10.1109/ICBSLP.2018.8554466

Scheerder, A., Van Deursen, A., & Van Dijk, J. (2017). Determinants of internet skills, uses and outcomes. A systematic review of the second- and third-level digital divide. *Telematics and Informatics*, *34*(8), 1607–1624. doi:10.1016/j.tele.2017.07.007

Scheinbaum, A. C. (2017). A framework for the dark side of social media : From Digital Drama to Digital Over-engagement. *The Dark Side of Social Media*, 3–10. doi:10.4324/9781315167718-1

Schenkman, L. (2021). Brain's sensory processor may prompt memory problems in autism. *Spectrum (Lexington, Ky.)*. doi:10.53053/INWE5797

Schütze, H., Manning, C. D., & Raghavan, P. (2008). *Introduction to information retrieval*. Cambridge University Press Cambridge.

Schwab, K. (2016). The Fourth Industrial Revolution: what it means and how to respond. *World Economic Forum*, 1–7.

SEDA. (2023a). *LSS – Renewable energy Malaysia*. SEDA Malaysia. https://www.seda.gov.my/reportal/large-scale-solar/

SEDA. (2023b). *Net Energy Metering (NEM) program*. SEDA.

Selvakumar, S., Adithe, S., Isaac, J. S., Pradhan, R., Venkatesh, V., & Sampath, B. (2023). A Study of the Printed Circuit Board (PCB) E-Waste Recycling Process. In Sustainable Approaches and Strategies for E-Waste Management and Utilization (pp. 159–184). IGI Global.

Senthil, T. S. R. Ohmsakthi vel, Puviyarasan, M., Babu, S. R., Surakasi, R., & Sampath, B. (2023). Industrial Robot-Integrated Fused Deposition Modelling for the 3D Printing Process. In Development, Properties, and Industrial Applications of 3D Printed Polymer Composites (pp. 188–210). IGI Global. doi:10.4018/978-1-6684-6009-2.ch011

Shafiabady, N., Lee, L. H., Rajkumar, R., Kallimani, V., Akram, N. A., & Isa, D. (2016). Using unsupervised clustering approach to train the Support Vector Machine for text classification. *Neurocomputing*, *211*, 4–10. doi:10.1016/j.neucom.2015.10.137

Shamsul Bahar, M. N. (2023). *CEO's message – Malaysian green technology and climate change Corporation*. Malaysian Green Technology And Climate Change Corporation – Leading Partner for Green Growth & Climate Change. https://www.mgtc.gov.my/who-we-are/ceos-message/

Shanta, S. S., Anwar, S. T., & Kabir, M. R. (2018, October 16). Bangla Sign Language Detection Using SIFT and CNN. *2018 9th International Conference on Computing, Communication and Networking Technologies, ICCCNT 2018*. IEEE. 10.1109/ICCCNT.2018.8493915

Shao, C., & Kwon, K. H. (2018). *Telematics and Informatics Clicks intended : An integrated model for nuanced social feedback system uses on Facebook. November*. Academia.

Sharma, P., & Singh, A. K. (2017). Association of state-wise food security index with climatic factors in India: Evidence from state-wise panel data. *Journal of Global Agriculture and Ecology*, *6*(3), 196–205.

Sheikhi, A., Rayati, M., Bahrami, S., & Ranjbar, A. M. (2015a). Integrated demand side management game in smart energy hubs. *IEEE Transactions on Smart Grid*, *6*(2), 675–683. doi:10.1109/TSG.2014.2377020

Shensa, A., Sidani, J. E., Escobar-Viera, C. G., Switzer, G. E., Primack, B. A., & Choukas-Bradley, S. (2020). Emotional support from social media and face-to-face relationships: Associations with depression risk among young adults. *Journal of Affective Disorders*, *260*, 38–44. doi:10.1016/j.jad.2019.08.092 PMID:31493637

Shepherd, C. (2011). *Viewpoint Does social media have a place in workplace learning ? 27*(2), 3–4. doi:10.1108/02580541111103882

Shoemaker, S. J., Wolf, M. S., & Brach, C. (2014). Development of the Patient Education Materials Assessment Tool (PEMAT): A new measure of understandability and actionability for print and audiovisual patient information. *Patient Education and Counseling*, *96*(3), 395–403. doi:10.1016/j.pec.2014.05.027 PMID:24973195

Siddiqua, F. (2014, August 22). The Silent Conversation. *The Daily Star*. https://www.thedailystar.net/the-silent-conversation-37929

Sievert, H., & Scholz, C. (2017). Engaging employees in (at least partly) disengaged companies. Results of an interview survey within about 500 German corporations on the growing importance of digital engagement via internal social media. *Public Relations Review*, *43*(5), 894–903. doi:10.1016/j.pubrev.2017.06.001

Singh, A. K., Kumar, S., & Jyoti, B. (2022c). Association of sustainable food security with climatic and non-climatic factors in Gujarat state of India: A district-wise panel data investigation. In: Leal Filho, W., Kovaleva, M., Popkova, E. (eds.), Sustainable Agriculture and Food Security. World Sustainability Series. Springer Cham. Springer Nature Switzerland. doi:10.1007/978-3-030-98617-9_16

Singh, A. K., Ahmad, M. M., & Sharma, P. (2017). Implications of socioeconomic factors on food security in selected economies: An empirical assessment. *Journal of Global Economics. Management and Business Research*, *8*(2), 103–115.

Singh, A. K., & Ashraf, S. N. (2020). Association of entrepreneurship ecosystem with economic growth in selected countries: An empirical exploration. *Journal of Entrepreneurship. Business Economics (Cleveland, Ohio)*, *8*(2), 36–92.

Singh, A. K., Ashraf, S. N., & Arya, A. (2019b). Estimating factors affecting technical efficiency in Indian manufacturing sector. *Eurasian Journal of Business and Economics*, *12*(24), 65–86. doi:10.17015/ejbe.2019.024.04

Singh, A. K., Ashraf, S. N., & Ashish, A. (2019c). A conceptual review on economic, business, intellectual property rights and science & technology related activities in Asian economies. [JJEM]. *JNNCE Journal of Engineering & Management*, *3*(2), 1–22.

Singh, A. K., Ashraf, S. N., Sharma, S. K., & Kumar, S. (2023e). Factors affecting farm income and climate adaptation strategies in the farming sector of Gujarat state of India. In B. J. Sadhik, J. Adewole, A. Yeneneh, & L. Abidoye (Eds.), *Opportunities and Challenges in Climate-Friendly Clean Water and Energy Technologies* (pp. 122–147). IGI Global. doi:10.4018/978-1-6684-7303-0.ch006

Singh, A. K., & Issac, J. (2018). Impact of climatic and non-climatic factors on sustainable livelihood security in Gujarat state of India: A statistical exploration. *Agriculture and Food Sciences Research*, *5*(1), 30–46. doi:10.20448/journal.512.2018.51.30.46

Singh, A. K., Issac, J., & Narayanan, K. G. S. (2019a). Measurement of environmental sustainability index and its association with socio-economic indicators in selected Asian economies: An empirical investigation. *International Journal of Environment and Sustainable Development*, *18*(1), 57–100. doi:10.1504/IJESD.2019.098641

Singh, A. K., & Jyoti, B. (2020). Factors affecting firm's annual turnover in selected manufacturing industries of India: An empirical study. *Business Perspectives and Research*, *2*(3), 33–59. doi:10.38157/business-perspective-review.v2i3.206

Singh, A. K., Jyoti, B., Kumar, S., & Lenka, S. K. (2021). Assessment of global sustainable development, environmental sustainability, economic development and social development index in selected economies. *International Journal of Sustainable Development and Planning*, *16*(1), 123–138. doi:10.18280/ijsdp.160113

Singh, A. K., & Kumar, S. (2021). Assessing the performance and factors affecting industrial development in Indian states: An empirical analysis. *Journal of Social Economics Research*, *8*(2), 135–154. doi:10.18488/journal.35.2021.82.135.154

Singh, A. K., & Kumar, S. (2022a). Factors influencing entrepreneurship ecosystem and economic development: Evidence from cross-sectional country-wise imbalanced panel data analysis. *The IUP Journal of Entrepreneurship Development*, *19*(3), 24–47.

Singh, A. K., & Kumar, S. (2022b). Exploring the impact of sustainable development on social-economic, and science and technological development in selected countries: A panel data analysis. *Society & Sustainability*, *4*(1), 55–83. doi:10.38157s.v4i1.405

Singh, A. K., Kumar, S., & Jyoti, B. (2022a). Influence of climate change on agricultural sustainability in India: A state-wise panel data analysis. *Asian Journal of Agriculture*, *6*(1), 15–27. doi:10.13057/asianjagric/g060103

Singh, A. K., Kumar, S., & Jyoti, B. (2023a). Does entrepreneurship ecosystem have a causal relationship with economic development? A country-wise imbalanced panel data investigation. In H. Magd, S. A. Khan, N. D. Nte, & M. Sarim (Eds.), *Entrepreneurship and Economic Sustainability*. Prestige Publishers.

Singh, A. K., Kumar, S., & Sharma, A. K. (2023c). Does green entrepreneurship ecosystem have causal association with economic development? evidence from country-wise panel data investigation. *Journal of International Business Research*, *22*(2), 1–18.

Singh, A. K., Kumar, S., Sharma, A. K., & Sinha, S. (2022b). Does green entrepreneurship have an association with sustainable development and its components? Evidence from a country-wise panel data investigation. In M. Magd, D. Singh, D. Spicer, & R. T. Syed (Eds.), *International Perspectives on Value Creation and Sustainability Through Social Entrepreneurship*. IGI Global USA. doi:10.4018/978-1-6684-4666-9.ch008

Singh, A. K., Sharma, A. K., & Jyoti, B. (2023d). Does economic development have a causal relationship with environmental degradation? Experience from different income group countries. In G. Sart (Ed.), *Considerations on Education for Economic, Social, and Environmental Sustainability* (pp. 300–333). IGI Global. doi:10.4018/978-1-6684-8356-5.ch015

Singh, A. K., Sharma, S. K., & Lenka, S. K. (2023b). Causality between green entrepreneurship and sustainable development: A cross-country analysis using ARDL model. *The IUP Journal of Applied Economics*, *22*(1), 1–34.

Singh, A. K., Singh, B. J., & Negi, V. (2020b). Does sustainable development have a causal relationship with environmental development? Evidence from a country-wise panel data analysis. *International Journal of Technology Management & Sustainable Development*, *19*(2), 147–171. doi:10.1386/tmsd_00020_1

Singh, A. K., Singh, B., & Ashraf, S. N. (2020a). Implications of intellectual property protection, and science and technological development in the manufacturing sector in selected economies. *Journal of Advocacy. Research in Education*, *7*(1), 16–35.

Sittón-Candanedo, I., Alonso, R. S., García, Ó., Muñoz, L., & Rodríguez-González, S. (2019). Edge computing, iot and social computing in smart energy scenarios. *Sensors (Basel)*, *19*(15), 3353. doi:10.339019153353 PMID:31370149

Smailhodzic, E., Hooijsma, W., Boonstra, A., & Langley, D. J. (2016). Social media use in healthcare: A systematic review of effects on patients and on their relationship with healthcare professionals. *BMC Health Services Research*, *16*(1), 1–14. doi:10.118612913-016-1691-0 PMID:27562728

Smith, J. K., & Johnson, M. A. (2022). Social media promotion of branded drugs. *Journal of Pharmaceutical Marketing*, *20*(3), 123–140.

Solomon, E. M., & van Klyton, A. (2020). The impact of digital technology usage on economic growth in Africa. *Utilities Policy*, *67*(10), 1–12. doi:10.1016/j.jup.2020.101104 PMID:32904493

Soomro, A. N., Kumar, J., & Kumari, J. (2022). The dynamic relationship between FDI, ICT, trade openness, and economic growth: Evidence from BRICS countries. *Journal of Asian Finance. Economics and Business*, *19*(2), 295–303.

Sperlich, W. B. (2006). *Noam Chomsky*. Reaktion Books.

Spring, M., Hughes, A., Mason, K., & McCaffrey, P. (2017). Creating the competitive edge: A new relationship between operations management and industrial policy. *Journal of Operations Management*, *49–51*(1), 6–19. doi:10.1016/j.jom.2016.12.003

Srinivasan, V. (2012). Multi generations in the workforce: Building collaboration. *IIMB Management Review*, *24*(1), 48–66. doi:10.1016/j.iimb.2012.01.004

Srite, M., & Karahanna, E. (2006). The role of espoused national cultural values in technology acceptance. *Management Information Systems Quarterly*, *30*(3), 679–704. doi:10.2307/25148745

Srivastava, S., Gangwar, A., Mishra, R., & Singh, S. (2022). Sign Language Recognition System using TensorFlow Object Detection API. *Communications in Computer and Information Science, 1534 CCIS*, 634–646. doi:10.1007/978-3-030-96040-7_48

Stanley, T. D., Doucouliagos, H., & Steel, P. (2018). Does ICT generate economic growth? A meta-regression analysis. *Journal of Economic Surveys*, *32*(3), 705–726. doi:10.1111/joes.12211

Statistics Austria. (2021). Urban-Rural-Typologie von Statistik Austria. STATatlas.

Strohmeier, S. (2007). Research in e-HRM: Review and implications. *Human Resource Management Review*, *17*(1), 19–37. doi:10.1016/j.hrmr.2006.11.002

Subha, S., Inbamalar, T. M., R, K. C., Suresh, L. R., Boopathi, S., & Alaskar, K. (2023, February). A Remote Health Care Monitoring system using internet of medical things ({IoMT}). *IEEE Explore*. doi:10.1109/ICIPTM57143.2023.10118103

Sun, Z., Sandoval, L., Crystal-Ornelas, R., Mousavi, S. M., Wang, J., Lin, C., Cristea, N., Tong, D., Carande, W. H., Ma, X., Rao, Y., Bednar, J. A., Tan, A., Wang, J., Purushotham, S., Gill, T. E., Chastang, J., Howard, D., Holt, B., & John, A. (2022). A review of Earth Artificial Intelligence. In Computers and Geosciences, 159. doi:10.1016/j.cageo.2022.105034

Sundar, S. S., & Limperos, A. M. (2013). *Uses and Grats 2.0: New Gratifications for New Media.* Taylor & Francis. doi:10.1080/08838151.2013.845827

Sun, W., Cai, Z., Li, Y., Liu, F., Fang, S., & Wang, G. (2018). Data processing and text mining technologies on electronic medical records: A review. *Journal of Healthcare Engineering*, *2018*, 2018. doi:10.1155/2018/4302425 PMID:29849998

Surianshah, S. (2021). Digital divide in education during COVID-19 pandemic. *Jurnal Ekonomi Malaysia*, *55*(3), 103–112. doi:10.17576/JEM-2021-5503-07

Svongoro, P. (2022). *The digital divide in the context of E-learning during the COVID-19 pandemic: A zimbabwean perspective.* doi:10.25159/UnisaRxiv/000041.v1

Syed, S., Robinson, J., & Reddington, B. (2022). Diagnostic dilemma in psychiatry: Disease of the mind or the media? *CNS Spectrums*, *27*(2), 251–252. doi:10.1017/S1092852922000645

Talei, H., Essaaidi, M., & Benhaddou, D. (2017). Smart campus energy management system: Advantages, architectures, and the impact of using cloud computing. *ACM International Conference Proceeding Series, Part F1305*, (pp. 1–7). IEEE. 10.1145/3128128.3128129

Tams, S., Ahuja, M., Thatcher, J., & Grover, V. (2020). Worker stress in the age of mobile technology: The combined effects of perceived interruption overload and worker control. *The Journal of Strategic Information Systems*, *29*(1), 101595. doi:10.1016/j.jsis.2020.101595

Tarafdar, M., Pirkkalainen, H., Salo, M., & Makkonen, M. (2020). Taking on the "Dark Side" - Coping with Technostress. *IT Professional*, *22*(6), 82–89. doi:10.1109/MITP.2020.2977343

Taylor, P. E., & Keeter, S. E. (2010). *Millennials: Confident. Connected. Open to Change*. Pew Research Center.

The Malaysian Industry-Government Group for High Technology (MIGHT). (2023). *MIGHT Portal*. Malaysian Industry-Government Group for High Technology. https://might.org.my/

Tiutiunyk, I., Drabek, J., Antoniuk, N., Navickas, V., & Rubanov, P. (2021). The impact of digital transformation on macroeconomic stability: Evidence from EU countries. *Journal of International Students*, *14*(3), 220–234. doi:10.14254/2071-8330.2021/14-3/14

Tiwari, S., Lane, M., & Alam, K. (2019). Do social networking sites build and maintain social capital online in rural communities? *Journal of Rural Studies*, *66*, 1–10. doi:10.1016/j.jrurstud.2019.01.029

Toivonen, H. (2022). Themes of climate change agency: A qualitative study on how people construct agency in relation to climate change. *Humanities & Social Sciences Communications*, *9*(1), 102. doi:10.105741599-022-01111-w

Turing, A. M., & Haugeland, J. (1950). *Computing machinery and intelligence*. MIT Press Cambridge. doi:10.1093/mind/LIX.236.433

Uddin, A., & Chowdhury, S. A. (2016, February 14). Hand sign language recognition for Bangla alphabet using Support Vector Machine. *2016 International Conference on Innovations in Science, Engineering and Technology (ICISET)*. IEEE. 10.1109/ICISET.2016.7856479

Valk, R., & Planojevic, G. (2021). Addressing the knowledge divide: Digital knowledge sharing and social learning of geographically dispersed employees during the COVID-19 pandemic. *Journal of Global Mobility*, *9*(4), 591–621. doi:10.1108/JGM-02-2021-0019

Van de Belt, T. H., Engelen, L. J., Berben, S. A., Teerenstra, S., Samsom, M., & Schoonhoven, L. (2013). Internet and social media for health-related information and communication in health care: Preferences of the Dutch general population. *Journal of Medical Internet Research*, *15*(10), e2607. doi:10.2196/jmir.2607 PMID:24088272

Van de WerfhorstH. G.KessenichE.GevenS. (2020). The digital divide in online education. Inequality in digital preparedness of students and schools before the start of the COVID-19 pandemic. doi:10.31235/osf.io/58d6posf.io/58d6p

Van de Werfhorst, H. G., Kessenich, E., & Geven, S. (2022). The digital divide in online education: Inequality in digital readiness of students and schools. *Computers and Education Online.*, *3*, 100100. doi:10.1016/j.caeo.2022.100100

Van Deursen, A. J., & Helsper, E. J. (2015). The third-level digital divide: Who benefits most from being online? *Communication and Information Technologies Annual*, 29-52. doi:10.1108/S2050-206020150000010002

Van Deursen, A. J., & Van Dijk, J. A. (2018). The first-level digital divide shifts from inequalities in physical access to inequalities in material access. *New Media & Society*, *21*(2), 354–375. doi:10.1177/1461444818797082 PMID:30886536

van Puijenbroek, T., Poell, R. F., Kroon, B., & Timmerman, V. (2014). The effect of social media use on work-related learning. *Journal of Computer Assisted Learning, 30*(2), 159–172. doi:10.1111/jcal.12037

Van Slyke, C., Bélanger, F., & Sridhar, V. (2005). A Comparison of American and Indian Consumers' Perceptions of Electronic Commerce. IGI Global. doi:10.4018/irmj.2005040102

van Uden-Kraan, C., Drossaert, C., Taal, E., Seydel, E., & van de Laar, M. (2008). Self-reported differences in empowerment between lurkers and posters in online patient support groups. *Journal of Medical Internet Research, 10*(2), e992. doi:10.2196/jmir.992 PMID:18653442

Vanitha, S. K. R., & Boopathi, S. (2023). Artificial Intelligence Techniques in Water Purification and Utilization. In *Human Agro-Energy Optimization for Business and Industry* (pp. 202–218). IGI Global. doi:10.4018/978-1-6684-4118-3.ch010

Vennila, T., Karuna, M. S., Srivastava, B. K., Venugopal, J., Surakasi, R., & B., S. (2023). New Strategies in Treatment and Enzymatic Processes. In *Human Agro-Energy Optimization for Business and Industry* (pp. 219–240). IGI Global. doi:10.4018/978-1-6684-4118-3.ch011

Verma, A., Dandgawhal, P. S., & Giri, A. K. (2023). Impact of ICT diffusion and financial development on economic growth in developing countries. *Journal of Economics, Finance and Administrative Service.* doi:10.1108/JEFAS-09-2021-0185

Verma, S., & Nidhi, V. (2017). *Extractive summarization using deep learning.* arXiv preprint arXiv:1708.04439.

Verma, S., Sharma, R., Deb, S., & Maitra, D. (2021). Artificial intelligence in marketing: Systematic review and future research direction. *International Journal of Information Management Data Insights, 1*(1), 100002. doi:10.1016/j.jjimei.2020.100002

Veveonah, M. (2020, June 16). *24 Hours on tree [Video].* YouTube. https://www.youtube.com/watch?v=-qxTyBWVLIU

Vignesh, S., Arulshri, K. P., Syedsajith, S., Kathiresan, S., Boopathi, S., & Dinesh Babu, P. (2018). Design and development of ornithopter and experimental analysis of flapping rate under various operating conditions. *Materials Today: Proceedings, 5*(11), 25185–25194. doi:10.1016/j.matpr.2018.10.320

Vijayarani, S., Ilamathi, M., & Nithya, C.(2015). Preprocessing techniques for text mining-an overview. *Sage, 5*(1), 7-16.

Vinod, G. (2021). Budget 2022: Increase allocation by RM2 bil to help East malaysians, non-malays. *Focus Malaysia.* https://focusmalaysia.my/budget-2022-increase-allocation-by-rm-2-bil-to-help-East-malaysians-non-malays/

Vyas, P., & Itam, U. J. (2022). *The Role of Social Media in Employee Knowledge Sharing.* IGI Global. doi:10.4018/978-1-6684-5849-5.ch009

Vyas, P. G., & Pandey, S. (2020). The effect of social networking sites use on employees' knowledge sharing. *European Journal of Training and Development*, *45*(2–3), 218–240. doi:10.1108/EJTD-02-2020-0027

Wadhawan, A., & Kumar, P. (2021). Sign Language Recognition Systems: A Decade Systematic Literature Review. *Archives of Computational Methods in Engineering*, *28*(3), 785–813. doi:10.100711831-019-09384-2

Wang, S., & Wu, P. (2008). *Computers & Education The role of feedback and self-efficacy on web-based learning : The social cognitive perspective*. Science Direct. doi:10.1016/j.compedu.2008.03.004

Wan, Z. W. Y., & Ismail, M. (2008). 기사 (Article) 와 안내문 (Information). *The Electronic Library*, *34*(1), 1–5.

Weiss, S. M., Indurkhya, N., & Zhang, T. (2015). *Fundamentals of predictive text mining*. Springer. doi:10.1007/978-1-4471-6750-1

Weizenbaum, J. (1966). ELIZA—A computer program for the study of natural language communication between man and machine. *Communications of the ACM*, *9*(1), 36–45. doi:10.1145/365153.365168

Weyland, A. (2011). Engagement and talent management of Gen Y. *Industrial and Commercial Training*, *43*(7), 439–445. doi:10.1108/00197851111171863

Whitacre, B. E., Alam, M. R., & Lobo, B. J. (2018). Econometric error nullifies finding of the impact of broadband speed on county-level employment. *Information Economics and Policy*, *44*, 58–60. doi:10.1016/j.infoecopol.2018.05.001

Whitacre, B. E., Gallardo, R., & Strover, S. (2014a). Does rural broadband impact jobs and income? Evidence from spatial and first-differenced regressions. *The Annals of Regional Science*, *53*(3), 649–670. doi:10.100700168-014-0637-x

Whitacre, B. E., Gallardo, R., & Strover, S. (2014b). Broadband's contribution to economic growth in rural areas: Moving towards a causal relationship. *Telecommunications Policy*, *38*(11), 1011–1023. doi:10.1016/j.telpol.2014.05.005

Whiting, A., Williams, D., Whiting, A., & Williams, D. (2013). *Why people use social media : a uses and gratifications approach*. Emerald Insight. doi:10.1108/QMR-06-2013-0041

Wicks, P., Massagli, M., Frost, J., Brownstein, C., Okun, S., Vaughan, T., Bradley, R., & Heywood, J. (2010). Sharing health data for better outcomes on PatientsLikeMe. *Journal of Medical Internet Research*, *12*(2), e1549. doi:10.2196/jmir.1549 PMID:20542858

Wirth, N. (2018). Hello marketing, what can artificial intelligence help you with? *International Journal of Market Research*, *60*(5), 435–438. doi:10.1177/1470785318776841

Wong, P. C., Whitney, P., & Thomas, J. (1999). Visualizing association rules for text mining. *Proceedings 1999 IEEE Symposium on Information Visualization (InfoVis' 99)*. IEEE.

Yahaya, N. (2020, September 10). KJ jumpa Veveonah, minta maaf wakili kerajaan. *Sinarharian.* https://www.sinarharian.com.my/article/100550/berita/nasional/kj-jumpa-veveonah-minta-maaf-wakili-kerajaan

Yan, A., Wen, S., Vogel, D., & Kwok, R. C. (2010). *Computers & Education Can learning be virtually boosted ? An investigation of online social networking impacts.* Science Direct. doi:10.1016/j.compedu.2010.06.015

Yang, Y. (2020). Use of herbal drugs to treat COVID-19 should be with caution. *Lancet, 395*(10238), 1689–1690. doi:10.1016/S0140-6736(20)31143-0 PMID:32422123

Yang, Y., & Liu, X. (1999). A re-examination of text categorization methods. *Proceedings of the 22nd annual international ACM SIGIR conference on Research and development in information retrieval.* ACM. 10.1145/312624.312647

Yap, M. Y. (2020, June 15). This Malaysian student spent 24 hours on a tree in order to get internet signal to complete her exams. *Mashable.* https://sea.mashable.com/culture/11036/this-malaysian-student-spent-24-hours-on-a-tree-in-order-to-get-wi-fi-signal-to-complete-her-exams

Yasir, F., Prasad, P. W. C., Alsadoon, A., & Elchouemi, A. (2016). SIFT based approach on Bangla sign language recognition. *2015 IEEE 8th International Workshop on Computational Intelligence and Applications, IWCIA 2015 - Proceedings*, (pp. 35–39). IEEE. 10.1109/IWCIA.2015.7449458

Yasir, F., Prasad, P. W. C., Alsadoon, A., Elchouemi, A., & Sreedharan, S. (2018). Bangla Sign Language recognition using convolutional neural network. *2017 International Conference on Intelligent Computing, Instrumentation and Control Technologies, ICICICT 2017, 2018-January*, (pp. 49–53). IEEE. 10.1109/ICICICT1.2017.8342533

Ye, J., Gao, Y., & Su, X. (2019). Smart Energy Management Cloud Platform Design Based on Offshore Wind Farm. *Proceedings - 2019 International Conference on Intelligent Transportation, Big Data and Smart City, ICITBS 2019.* IEEE, 10.1109/ICITBS.2019.00038

Yigit, M., Gungor, V. C., & Baktir, S. (2014). Cloud Computing for Smart Grid applications. *Computer Networks, 70*, 312–329. doi:10.1016/j.comnet.2014.06.007

Zaccai, E. (2012). Over two decades in pursuit of sustainable development: Influence, transformations, limits. *Environmental Development, 1*(1), 79–90. doi:10.1016/j.envdev.2011.11.002

Zhai, C., & Lafferty, J. (2017). A study of smoothing methods for language models applied to ad hoc information retrieval. *ACM SIGIR Forum.* ACM New York, NY, USA. 10.1145/3130348.3130377

Zhang, R., Cheng, X., & Yang, L. (2016). Energy Management Framework for Electric Vehicles in the Smart Grid: A Three-Party Game. *IEEE Communications Magazine, 54*(12), 93–101. doi:10.1109/MCOM.2016.1600335CM

Zhang, Zhao, S., Wan, X., & Yao, Y. (2021). Zhang, W., Zhao, S., Wan, X, & Yao, Y. (2021). Study on the effect of digital economy on high-quality economic development in China. *PLoS One, 16*(9), e0257365. doi:10.1371/journal.pone.0257365 PMID:34547019

Zhao, Y., & Zhang, J. (2017). Consumer health information seeking in social media: A literature review. *Health Information and Libraries Journal, 34*(4), 268–283. doi:10.1111/hir.12192 PMID:29045011

About the Contributors

Patricia Ordóñez de Pablos is a professor in the Department of Business Administration in the Faculty of Business and Economics at The University of Oviedo, Spain. She completed her education in The London School of Economics, UK. Her teaching and research interests focus on the areas of strategic management, knowledge management, organizational learning, human resource management, intellectual capital, information technologies, with special interest in Asia (Bhutan, China, Laos, Myanmar). She is Editor-in-Chief of the International Journal of Learning and Intellectual Capital (IJLIC) and International Journal of Asian Business and Information Management (IJABIM), respectively. She has edited books for IGI Global, Elsevier, Routledge, and Springer. In 2021, she earned placement on Stanford University's "Ranking of the World Scientists: World's Top 2% Scientists" list.

***

Manimuthu Ayyannan has over 12 years of experience in Software Engineering. He has served as a leader, architect, and engineer in various domains, including Distributed Systems, Big Data, Machine Learning, Artificial Intelligence, and Data Science. Manimuthu has a strong background in designing and building data-intensive complex systems, platforms, infrastructures, and enterprise applications. He is proficient in both object-oriented and functional programming languages and has expertise in utilizing multiple cloud technologies, such as GCP and Azure. Manimuthu Ayyannan is a renowned keynote speaker, having delivered speeches at numerous internal and external conferences, including the International Data Engineering and Science Association and Data Stack Summit. He has contributed to the field by presenting several papers at various international conferences. Additionally, he has authored or co-authored a couple of textbooks (ISBN:9789357571258, ISBN:9789357571487) on relevant subjects. Manimuthu actively participates in professional communities, being a member of the ACM and IEEE communities. He is associated with the IEEE Computer Society, IEEE Computer Society Technical Community on Data Engineering, and IEEE Big Data Community, among others.

Sampath Boopathi completed his undergraduate in Mechanical Engineering and postgraduate in the field of Computer-Aided Design. He completed his Ph.D. from Anna University and his field of research includes Manufacturing and optimization. He published 100 more research articles in Internationally Peer-reviewed journals, one Patent grant, and three published patents.He has 16 more years of academic and research experiences in the various Engineering Colleges in Tamilnadu, India. ().

Sahil Manoj Chavan is a faculty member in the Department of Electrical Power System at Sandip University, located in Nashik, Maharashtra 422213, India

Maureen Hassall, is the Director of the Minerals Industry Safety and Health Centre and the UQ R!SK research initiative. She has over 30 years' experience working in industry and in academia developing and leading innovative transformational change projects in the areas of risk, human factors, process safety and systems safety engineering.She is a registered and chartered professional engineer and a certified practicing ergonomist.

Ajeya Jha is a professor at the Department of Management Studies at Sikkim Manipal Institute of Technology (SMIT) since February 2000. In addition to having post-graduate and doctoral degrees in Business Administration, he possesses graduation and post graduation in Pharmaceutical Sciences also. fellow of International Society of Management. Has a rich and diverse industrial experience of 15 years (includes launching two green-field projects), teaching 21 years and with NGO 8 months. He is the founder Chief coordinator of Entrepreneurship Development Cell (EDC) at SMIT established with support from Department of Science and Technology, Government of India.

Bhim Jyoti has been working with V.C.S.G., UUHF, College of Forestry, Ranichauri, Tehri Garhwal, Uttarakhand, India as Assistant Professor of Seed and Technology. She received MSc in Seed and Technology from CCS University Meerut and PhD in Seed and Technology from G. B. Pant University of Agriculture, Pantnagar, Uttarakhand. She has published several research articles in the seed and technology, plant growth, seed vigour, genetic variability, Biotechnological approaches, climate change and agricultural productivity.

Tajbia Karim is currently pursuing her PhD in Center for Artificial Intelligence Technology (CAIT), Faculty of Information Science and Technology, Universiti Kebangsaan Malaysia (UKM). She is employed as an Assistant Professor in the Department of Information and Communication Tchnology in Bangladesh University of Professionals (BUP). Her research interests are Computer Vision, Image Processing, Machine Learning and Augmented Reality.

Chen Jung Ku is currently a Mandarin language teacher of Universiti Malaysia Sabah.

Jitendra Kumar is currently working as an Associate Professor in Department of Management Studies, at Sikkim Manipal institute of Technology. He has completed his post graduation in Master of Business Administration from Sikkim Maniapl University, Gangtok Sikkim in the year 2008. He is having 10 years of teaching experience in finance domain at college level. His area of interest are in Finance, Financial market and investment. He got his doctorate degree from Sikkim Manipal University, India in the year 2016.

Yoke Lian Lau currently is a foreign language lecturer from Universiti Malaysia Sabah.

Chek Kim Loi is a senior lecturer at the Universiti Malaysia Sabah. She obtained her PhD from the University of Otago, New Zealand.

Zainal Rasyid Mahayuddin is a Senior Lecturer in the Center for Artificial Intelligence Technology (CAIT), Faculty of Information Science and Technology, Universiti Kebangsaan Malaysia (UKM). He specializes in Simulation Modelling, Visual Computing, Mixed Reality (Virtual and Augmented Reality) and Computer Vision.

Samrat Mukherjee has been working as a faculty for last 11 years, and has published more than 20 papers in different journals.

Subhadeep Mukherjee is currently associated with at Department of Management Studies, Ramaiah University of Applied Sciences . He is also a former faculty of Department of Management Studies, National Institute of Technology Silchar and also having Industry Experience in the domain of HR and Marketing for many years. He has published many papers in Scopus, ABDC, Web of Science and UGC Indexed reputed journals and also presented many papers in International and National Conferences. He is performing the responsibility as Co-Editor – in –Chief of Adhyatma: A Journal of Management, Spirituality and Human Values and also as the reviewer of many International and National Scopus Indexed Journals. He is also associated as Life Member with many esteemed organisation like Indian Commerce Association, CEGR (Centre for Education Growth and Research), New Delhi,

Social Science Research Network [SSRN], U.S.A, Social Science and Humanities Research Association (SSHRA) – ICBELLP.

Ernahwatikah Binti Nasir is a Mandarin language teacher of Universiti Malaysia Sabah.

P.V. Pramila is a faculty member in the Department of Electrical and Electronics Engineering at Hindusthan College of Engineering and Technology, situated in Valley Campus, Coimbatore, Tamil Nadu 641032, India.

Nikodem Rybak is a machine learning, natural language processing and complex systems research engineer at the University of Queensland. His main area of research is AI-driven problem solving for complex systems including language, human economies and political systems, climate, nervous systems, cells and organisms, as well as modern information, energy and telecommunication infrastructures.

Abhijit Sarkar is an Associate Professor and Head of the Research Proposal Evaluation Committee at the School of Management at SRM University, Sikkim. He is an MBA from ICFAI Business School and has earned his doctoral degree from Sikkim Manipal Institute of Technology (SMU). His keen interest in Road Freight Transport particularly in terms of Heavy Trucks has led him to pursue research on Heavy Trucks (Multi – Axle Trucks) as a part of his doctoral research. He has several research publications (in peer reviewed journals) in the area of Indian Road freight transport industry, entrepreneurship, services sector and retail industry. His area of research interests and expertise are road freight transportation and branding related aspects among unorganized small to medium sized fleet operators and their value perceptions and preferences.

Ajay K. Singh is working as an Assistant Professor (Economics) in the School of Liberal Arts & Management, DIT University Dehradun. He did Post-Doctorate Research with EDI of India, Ahmedabad, Gujarat (India). He received MPhil (Economics) from Devi Ahilya University Indore (India), and PhD (Economics) from IIT Indore (India). He has published research papers in the area of climate change and agricultural productivity, assessment of food security; estimation of GFSI, assessment of environmental sustainability index and its association with socio-economic indicators; implications of entrepreneurship ecosystem on economic development; and implications of digitalization on economic development at national and global level.

Narayanaswamy Venkateswaran is working as Professor at Panimalar Engineering College, having 26 years of teaching experience published more papers in

International journals and also member of editorial board in more reputed journals. Published books on International Business Management, HR Analytics, Entrepreneurship Development, Professional Ethics and Human Values, Fundamentals of Management. Area of expertise is on Business Analytics, Artificial Intelligence and Supply Chain Management"

I K Vidhya is a faculty member in the Department of Computer Science and Engineering at Karunya Institute of Technology and Sciences, situated in Coimbatore, Tamil Nadu 641114, India.

Jenny Tze Kie Yap currently is a Mandarin lecturer of UiTM, Sarawak.

Zi Hong Yong currently a Mandarin language teacher of Universiti Malaysia Sabah.

Index

K

L

M

N

O

P

Q

S

T

U

V

W

www.ingramcontent.com/pod-product-compliance
Lightning Source LLC
LaVergne TN
LVHW061537290926
849162LV00004B/819